A TRIP OF ONE'S OWN

HOPE, HEARTBREAK, AND WHY TRAVELING SOLO COULD CHANGE YOUR LIFE

Kate Wills

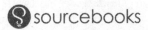

Published by Sourcebooks
P.O. Box 4410, Naperville, Illinois 60567-4410
(630) 961-3900
sourcebooks.com

Originally published as *A Trip of One's Own* in 2021 in the United
Kingdom by Blink Publishing, an imprint of Bonnier Books UK.

Library of Congress Cataloging-in-Publication Data

Names: Wills, Kate, author.
Title: A trip of one's own : hope, heartbreak, and why traveling solo could
    change your life / Kate Wills.
Description: Naperville : Sourcebooks, [2022] | "Originally published as A
    Trip of One's Own in 2021 in the United Kingdom by Blink Publishing, an
    imprint of Bonnier Books UK" -- Title page verso. | Includes
    bibliographical references.
Identifiers: LCCN 2021037094 (print) | LCCN 2021037095 (ebook) |
    (trade paperback) | (epub)
Subjects: LCSH: Wills, Kate. | Travel writers--Biography. | Women
    travelers--Biography. | Travel. | Travel journalism.
Classification: LCC G154.5.W55 A3 2022 (print) | LCC G154.5.W55 (ebook) |
    DDC 910.4092 [B]--dc23/eng/20211110
LC record available at https://lccn.loc.gov/2021037094
LC ebook record available at https://lccn.loc.gov/2021037095

Printed and bound in Canada.
MBP 10 9 8 7 6 5 4 3 2 1

For Julia, who took me on my greatest adventure, and for Blake, whose adventures have only just begun.

"To awaken quite alone in a strange town is one of the pleasantest sensations in the world."

Freya Stark

CONTENTS

Arrivals   1

Sister Act   15
Sleepless in Shanghai   32
On Assignment   50
The Women's Movement   66
A Man's World?   82
Fight or Flight?   97
The Traveling Cure   112
Footloose and Fancy-Free   121
The Great Escape   140
Life in the Fast Lane   155
Destination Unknown   167
Home Sweet Roam   182

Departures   197
Reading Group Guide   203
A Conversation with the Author   205
Bibliography   208

Acknowledgments   212
About the Author   214

ARRIVALS

"You are traveling alone?" asked the Israeli border guard, eyeing up my suspiciously small carry-on luggage and disheveled appearance. If the severity of the way she applied her lipliner was anything to go by, I really didn't want to piss her off.

I hesitated, even though it's a question I've been asked countless times. I was *sola* in Mexico City, *unico* in Rome, and *mongwe* in Botswana. I'd worn a fake wedding ring in India. (Supposedly to deter unwanted male advances. It didn't work.) I'd worn a real wedding ring in Amsterdam and Arizona. But this was the first time I'd ever really flown solo, solo.

Although I'd been a travel journalist for over ten years and was used to jetting in and out of unfamiliar cities with only my laptop for company, I'd never really felt like I was going it alone. There would often be a photographer in tow or other journalists—in a particularly strange circus known as "the group press trip." Even when I did embark on an adventure by myself—three months of volunteering in India where I hoped that I would somehow "find myself"; a stint living in LA, which I'd called "a marriage sabbatical"—I knew that my partner, Sam, was back at home, the patient Penelope to my Odysseus. If something exciting happened, he was the first person I'd text. If I'd had a tough day, he'd be there on the other end of the phone to make it all better. At times I felt like he was my true north, the grounding force I always came home to. It was only now that I found myself without him that I realized how much I had relied on him. The reason I had been able to travel so far and for so long was because I had felt the strength of his support back at home.

But now we were getting divorced. Even the word sounded horrible. I didn't know anyone else who was getting divorced, and certainly not in their early thirties. I felt as if I'd been prematurely pushed into a more mature age bracket, like going through early menopause. Well-meaning friends tried to sympathize: "When my seven-year relationship ended…" they'd begin, not understanding the unique pain I felt. To have declared, in front of everyone you love and respect most in the world, that you will spend your life with a person, only to then spectacularly fail, is a shockingly singular experience.

I was now officially single. For the first time since I was twenty-one, I wasn't leaving anyone behind when I jetted off on yet another adventure. No one was going to miss me, or so it felt. Although my friends and my sister had been amazing—turning up with pizza that I was too sad to eat, texting me hourly to check that I was OK—they all had their own lives and families. I was thirty-four, and I felt completely and utterly alone.

As a serial monogamist, I had been used to always having a "someone." Someone to visualize in your head when you hear a love song on the radio. Someone to daydream about bringing back to the amazing place you'd just discovered. Someone to show your tan line off to when you got home. But I didn't even have a home anymore. Following the breakdown of my marriage—and then a passionate rebound love affair with my friend Guy that ended in further devastating heartbreak—I'd rented out my flat and put all my worldly possessions into boxes, which were now shoved into my friend Josh's spare room. So far, so *Eat, Pray, Love*.

Whenever I meet people while traveling solo, the most common comment is, "You're brave." It was similar when I told people I was getting divorced. The truth is that I've never felt particularly brave while traveling on my own. I've felt stupid, disorientated, and embarrassingly ill-equipped (like the time I tried to hike the foothills of the Himalayas in flip-flops) but never really brave. Bravery is when you're scared of something, but you do it anyway. Traveling for me isn't scary. It can be hard, but most of the time it's too rewarding and exhilarating to dwell on the fact that you're doing it solo. But going through life alone…as I was now? That felt truly terrifying.

There must have been a moment when I realized that my choices would result in the immediate destruction of everything in my life, but it's hard to pinpoint when that was. It was about a year ago when the nagging buzz that something wasn't right became more of a roar. Up until that point, I'd told myself it couldn't possibly be my relationship of thirteen years. We had reclaimed wood parquet floors and a joint Tate membership and had co-created a world together with all its inside jokes and nicknames and silly songs and pretending to be a ghost whenever we changed the duvet cover.

And yet something had to give. I regularly found myself crying in the shower. Lather, rinse, re-weep. I'd made some big changes in my life to try and make the niggle of not-quite-rightness go away. I had quit my dream job on a national newspaper and gone freelance (another "you're brave" moment that didn't feel brave, just reckless). I embarked on what would become five years of psychoanalysis—hoping that this intense form of therapy where you lie on a couch four times a week could help unpick why I felt completely numb, like I was underwater all the time. I went on the aforementioned soul-searching pilgrimage to India. Nothing worked.

During all these changes, I convinced myself that there are two types of people in life—those who gingerly lower themselves into a swimming pool and those who dive in. After a lifetime of only dipping a toe, I felt I had to cannonball into something. I thought perhaps taking the plunge and getting married might be what we needed to make things better. So at aged thirty-two, I used the impetus of the Leap Day in February to jump right in and propose to Sam on Hampstead Heath. Any doubts that surfaced I put down to cold feet or the stress of "wedmin." The wedding juggernaut was set in motion. I felt buoyed by the excitement of choosing dresses and booking a venue, and when the bad feelings came, I pushed them away by continuously taking more and more trips by myself. Travel became a way to forget myself for a little while, in the way that some people use sex or drugs or exercise.

I got married for many reasons. I liked the way he saw me better than I liked myself. I had loved him for so long, and I felt more real with him than I ever had with anyone else. He challenged me and made me feel smarter just by being in the same room as him. He read me poetry every day, and his eyes welled up with tears when he listened to songs from musicals. He was a Good Person in a way that I was not. No doubt I also got married to feel more grown-up and because everyone else was. These reasons seemed no better or worse than others I'd heard.

Our wedding was genuinely the best wedding I've ever been to (and I'm not just saying that because it was mine). The honeymoon was equally blissful. But when we got back, I realized that the doubts and sadness I was feeling were significantly more than just post-wedding blues. I had been thinking that getting married would mean everything would change, but nothing had changed. All the problems we had before were still there, but with the added pressure of "forever" weighing down on us. I realized that for a long time, we had gradually been drifting apart, but so slowly and incrementally, like tectonic plates, that neither of us had noticed. Staring down the barrel of "the rest of our lives," it dawned on me that we had completely different pictures of what that actually looked like. To paraphrase E. B. White, he wanted to save the world, but I wanted to savor it.

When I had to tell Sam I didn't love him anymore—one brazenly sunny morning after I'd lain awake all night—I felt like I'd thrown a grenade into the room. To him, it must have seemed completely without warning. I was wracked with guilt at being the one who wanted to end things but convinced that he deserved to be with someone who was in love with him and who wanted the same kind of life. Still, it felt like a tsunami of pain and grief and unbearable change that threatened to sweep me away.

We tried counseling, but I had made up my mind. It was as if with this one decision, lots of other things in my life that had felt wrong now made sense. On our first wedding anniversary, we were living separately. And now, nine months later, I was overwhelmed with sadness and still trying to make sense of the fallout.

•———→

"Yes, it's just me," I told the Israeli border guard, wondering why I always add the "just." Because having no plus-one is a great way to see the world. I'd been extolling the virtues of having solo adventures for years and had written many articles on the rise of this traveling trend. I knew that 27 percent of people now take solo breaks, compared to just 10 percent a decade ago, and 55 percent of solo travel searches in the United Kingdom are made by women. I'd spoken to trend forecasters who told me that going on holiday by yourself was no longer perceived as the preserve of sad loners paying a single supplement but a sign of female wealth, independence, and freedom.

So why did I suddenly feel so lonely? There's something about airports that has always inspired existential dread in me—a sense of malaise that not even duty-free Daim bars can make up for. Maybe it's the eye-wincingly bright lights, the too-shiny floors, or the fact that you're usually sleep deprived. Or is it the sense of being anywhere and nowhere, with hours of time to pass, that makes it all seem meaningless? If sociologist Jean Baudrillard thought shopping malls were the epitome of hyperreality, he should've spent more time in airports. I don't think I'm alone in being reminded of my own mortality while waiting for my gate number and sitting in a sadistically uncomfortable plastic chair. There's a reason they're called terminals.

And yet airports are a space where it's decidedly okay to be alone—appropriate, even. Yes, there are loved-up couples going on getaways and sash-wearing bachelorettes sipping prosecco. But there are also plenty of just-ones without plus-ones. When I'd left London, I'd watched people line up at security clutching their passports and visas, removing their shoes and sealing up their dreams along with their toothpaste into tiny plastic bags, all obediently going through the rigamarole that is modern travel, and I wondered not for the first time where this collective urge to wander comes from. Why do we endure the indignity and annoyances to board a giant metal tube that takes us into the upper troposphere, so we can breathe recycled air and eat non-food off plastic trays? Why traipse from point A to point B at all?

To travel for pleasure is an odd impulse, and it's a relatively recent one. In Yuval Noah Harari's *Sapiens*, he suggests that going on holiday is a way of buying into the myth of romantic consumerism, that "to make the most of our human potential we must have as many different experiences as we can… One of the best ways to do all that is to break free from our daily routine, leave behind our familiar setting, and go traveling in distant lands, where we can 'experience' the culture, the smells, the tastes, and the norms of other people." He points out that a wealthy man in ancient Egypt would never have dreamed of solving a relationship crisis by taking his wife on a getaway to Babylon.

It was the Romans who pioneered the idea of traveling for fun. They even had guidebooks listing places of interest and the distances between them—Pausanias's *Description of Greece* in the second century being the first known example. But with the fall of Rome, visiting other lands fell out of favor. The constant threat of battle and unsafe travel routes meant that most people didn't venture far beyond their neighboring village if they didn't need to. The exception was the religious pilgrimage, and the word "holiday" is derived from the Old English word *haligdaeg*, meaning a "holy day" or saint's day in the Christian calendar.

During Tudor times, travel was mainly reserved for kings and queens on the "royal progress," when the monarch went on tour with their entourage to display their pomp and power to the provinces. On one occasion, Henry VIII took 4,000 people with him, like the biggest bachelor party ever. As trade and exploration increased during the Renaissance, so too did overseas adventures. Hernán Cortés, Christopher Columbus, and Marco Polo were out circumnavigating and, for better or worse, carving up the globe on behalf of European powers and changing the course of history.

In the seventeenth and eighteenth centuries, rich young gentlemen began taking a "Grand Tour" of Europe for several months, usually culminating in Rome. Although at first it was only the boys who would swan off for self-improving cultural tours, eventually their sisters began to get in on the act, too. Of course, these young women would have to be accompanied by a chaperone—normally a spinster aunt, who would ideally be given the slip as soon as possible.

Travel for the sake of curiosity became a prized pursuit for the cultivated person, and the mind-expanding benefits of being elsewhere started to be understood. These trips resemble the holiday as we know it today. Grand Tour-ists visited historical sites, viewed works of art and architecture, practiced their language skills, caught up with Continental fashion, took lessons in dancing or fencing, and collected souvenirs that could later be used to show off how cultured they were. The gap year was born.

The Industrial Revolution saw a boom in a variety of mass transport, from bicycles to railways and ocean liners, which enabled independent travel by men *and* women, and not just the upper echelons of society. In 1841, Thomas Cook arranged his first tour, a bespoke train trip between Leicester and Loughborough. With the introduction of the bank holiday in 1871, British pleasure-seekers could jump on a train and spend a three-day break in one of the many "sea-bathing" resorts along the coast.

The first commercial flight in 1928 (the German Graf Zeppelin), the introduction of two weeks' paid holiday, and the post-war economic boom made going abroad more mainstream. In 1950, one million Brits traveled overseas, many of them on these new-fangled things called *package holidays*. In 1973, a little guide book to Asia published by Lonely Planet came out and by the 1990s low-budget airlines were soaring. In a single weekend in July 2019, two million people flew out of UK airports. That record figure was expected to be surpassed in 2020 until…well, we all know what happened.

The freedom to travel is a privilege. You must be settled and safe to desire this kind of mobility. You must have free time, a passport, disposable income, and a body that allows it. Travel does not mean the same thing to me as it does to an asylum seeker, crossing continents with their worldly possessions in a single bag, risking their life in search of a better one.

What does it mean to be a traveler and not a tourist? Are we going on holiday or are we having an adventure? The word *adventure* comes from the Latin word *adventurus,* meaning "about to happen," which sums up the thing I love best about being away from home. The feeling that something exciting might be just around the next corner. Even if that thing is only an average Airbnb or a bout of food poisoning.

The word *adventure* has associations of chance and luck and the best trips feel like this, too. Anyone who has felt the thrill of just scraping onto a flight after final call, or of happening to catch a brilliant street performer on a random street, or of stumbling across the best restaurant in town (steamed-up windows, packed with locals, and, miraculously, with just one free table, seemingly waiting just for you), knows that there's always an element of chance and risk in the best travel experiences.

Adventures don't have to be in exotic locales. It is possible to get this same feeling by being really badly organized and not planning anything at all—especially if you tell a good enough story about it afterward. For Jean-Paul Sartre, adventures only existed in the retelling: "For the most banal even to become an adventure, you must (and this is enough) begin to recount it." I'm pretty sure you could have an adventure in Slough if you just rocked up at the station with five pounds to your name and tried to get home. By the fourteenth century, the word *adventure* took on meanings of danger. Which is probably also on offer in Slough with a fiver in your pocket.

It's often said that travel broadens the mind, and it turns out there might be some truth in this. An experiment at Indiana University divided students into two groups, and both were asked to list as many different modes of transportation as possible. One group was told that the task was developed by Indiana University students studying abroad in Greece (the distant condition) while the other group was told that the task was developed by Indiana University students studying in Indiana (the near condition).

Researchers found a striking difference between the two groups. When students were told that the task was imported from Greece, they came up with so many possibilities. They didn't just list buses, trains, and planes, they cited horses, spaceships, bicycles, and Segway scooters. Because the source of the problem was far away, the subjects didn't just think about getting around in Indiana—they thought about getting around all over the world and even in deep space. When we untether our mind from routine and habit and give it something new to chew on, new connections are made. It can also offer up some much-needed perspective. Perhaps the French novelist Gustave Flaubert put it best: "Travel makes one modest; you see what a tiny place you occupy in the world."

Considering I've made my living as a travel writer, I've never actually been very good at the traveling part. As a child, I got travel sick and barely left Surbiton. One memorable summer, my family drove all the way to the south of France, and I spent the whole journey projectile vomiting in the back of the car. It wasn't exactly *A Year in Provence*, although it felt like it took that long to get there. After my parents split up, my mum would often take me and my sister on holidays. But they were never to places we'd want to go to, and when we got there she would often sleep all day as we watched dubbed TV shows in random apartments in bleached-out, run-down beach resorts.

Aged seventeen, I did the obligatory "Girls on Tour" holiday to Malia in Crete, but I still felt very unworldly when I got to university, and it seemed like every girl in my class was called Portia or Pandora and had just got back from Peru. I made up for this by backpacking around Thailand during the summer holidays and going on a study abroad trip when I was twenty. But while other students on the scheme opted to go to universities in culturally improving destinations where they would also learn a language, I picked UC Santa Barbara, primarily because I loved *The OC*.

It was actually Sam—I'm slightly annoyed to admit—who opened my mind to the possibilities of more adventurous travel when I was twenty-two. Having finished university, he suggested we "spend a few months in Guatemala," as if it were the same as going to Guildford. I was terrified. But like everyone else I'd met at university, he'd had a "gap yah" and, as my plans for the summer had involved mailing out CVs for journalism internships I had little chance of getting, off to Central America we went.

It was here, on the shores of Lake Atitlán, that I finally got the travel bug. Sadly that travel bug was amoebic dysentery. Or at least, I always like to claim that it was. All intrepid travel journalists need at least one exotic illness under their belt. And that's what an old man in my Spanish class diagnosed me with after more than seven days when I could barely keep water down and had a fever so high I was hallucinating patterns on the ceiling.

Sam and I were staying with a lovely Guatemalan family at the time and the only toilet was a cockroach-infested hole in the ground with a pink plastic jug of fetid water to "flush" it, accessed by creeping through the children's bedroom. I can honestly say it's the sickest I've ever been. Most nights I was delirious, feeling like I'd crossed a barrier between worlds. We were so far from a hospital that making it out of Central America alive seemed an impossibility.

But eventually I recovered, though things only got more, er, adventurous from there. On that trip we were spat at in the street in El Salvador and had all our belongings—including our glasses and contact lenses—stolen on an overnight bus in Nicaragua, meaning we were not only broke and hungry, but blind too. Although by most people's standards that trip was pretty shambolic, it was my first real adventure. I had that feeling for the first time—one which I have subsequently been chasing like any good addict ever since—of being both completely lost and also totally free.

When I landed my first job in journalism, I began dedicating my career not to the pursuit of truth or giving a voice to the voiceless but to the visceral thrill of simply being somewhere new. When I realized that I could get paid to travel, there was no stopping me. I went on every press trip available, whether it was interviewing DJs in Ibiza or vampire-hunting in Transylvania. I became obsessed with putting some distance between myself and everything I knew—both the monotony of everyday life and the big soul-searching questions that would creep in from time to time. It was as if I thought that the meaning to life could be found in a well-packed suitcase and a freshly printed boarding pass.

During my twenties, travel seemed like an easy way to make myself more interesting. I came from the suburbs; I had mousey hair. I was average at most things. I was medium. I was beige. I was completely middle-brow. People who traveled seemed impossibly glamorous. They wore cashmere wraps in business class. They sipped coffee from takeaway cups as they briskly made their way through departure lounges, and they complained about jet lag. They brought back wicker bags and kooky ornaments and when you complimented them on their dress, they said they picked it up from a Mexican flea market for ten pesos.

Travel became my passport to having a personality. As I hurtled toward my thirties, going away all the time became a convenient way to dodge the "When are you getting married?" and the "Are you going to have a baby?" questions. For most of the people around me, it seemed like settling down was the goal, but I wanted to pack up and leave at every opportunity.

I kept waiting for my maternal gene to kick in, but backpacks still excited me more than onesies. I willed myself to look at friends' little darlings with something approximating desire for one of my own, but I just couldn't muster it. I'd listen patiently to stories about mastitis and fourth-degree tears and file it away in my brain for possible later use, but even the sweetest of babies held zero interest to me. Sometimes under duress, I'd hold friends' offspring for a bit but then my arms would ache, and I'd resent not having both hands free for snacking.

Travel was also an instant answer to the dreaded "What have you been up to?" question. In this way, travel is a great leveler. Holidays cross class, race, and cultural boundaries like nothing else (except maybe weddings and kids, but I wasn't very interested in those). There's a reason why the classic hairdresser line is, "You going anywhere nice on your holidays?"

If Joan Didion was right, and we tell ourselves stories in order to live, then a travel story is the best story of them all. And I had experience escaping in stories. Until I could travel, books were my way of going places. Growing up, my family life had been chaotic. Behind the veneer of middle-class respectability, there were police showing up at the door to say my dad had crashed the family car while drunk, and my mum was so neglectful that I once got into a boiling hot bath and ended up with blisters all over my feet. When their fights became too much, their behavior too erratic, I did have one constant I could always escape to, which is probably the reason I'd read all the Brontë sisters' books by the age of twelve.

No doubt the stories I read of female travel and adventure inspired my own globe-trotting. From Alice's madcap tumble down the rabbit hole in *Alice's Adventures in Wonderland* to the Wife of Bath's rabble-rousing ramble to Canterbury. From Clarissa striding across London in *Mrs. Dalloway* to Miss Quested's feverish mishaps in *A Passage to India*. I admired ballsy Miss Eleanor Lavish in *A Room with a View*, throwing away Lucy's guidebook in order to experience "the true Italy." And my most beloved of all, Dorothy's circular trip to Oz.

———▶

Now that my life as I knew it was in ruins, and I was *still* trying to get away from it all, I started to wonder about the women who had gone before me. The ones who had also felt this need to shuttle across foreign lands, miles from home. Marco Polo, Captain Cook, Bruce Chatwin, Jack Kerouac, and Paul Theroux are all household names, but where were their female counterparts? Why do we romanticize the solo male traveler—think of the lonesome cowboy striding into the sunset or the "Born to Run" hero hitting the open road—but a woman traveling by herself is often seen as tragic or transgressive? Who were the intrepid female travel writers? I realized I couldn't name a single one, really. There was the American pilot Amelia Earhart. I vaguely knew of Gertrude Bell, some nineteenth-century lady who did something important in the Middle East. But that was about it. So, I went back to the beginning.

After googling "first female travel writer" (obviously), I was amazed to find that one of the first ever travel writers was a woman. A nun named Egeria. All the way back in the fourth century. A time so far away from solo travel as I knew it, beyond baggage allowances and boarding gates and booking websites, that it was impossible to imagine what her journey may have been like.

As I began to delve into the subject more, I realized that there were many women like Egeria, blazing a trail. Why had their stories been pushed to the margins, forgotten and overlooked? Where did they find the energy to flout expectations of the "weaker sex," accept the risks of rape and kidnap and violence, and travel so far? How did they negotiate the self-doubt and the anxiety of looking beyond their horizons, of think-ing outside their role as keeper of the hearth and home? How did they know that travel was not just a male prerogative and that the world was also theirs for the taking? Why did they wander off despite the constant pressure to sit down and stay quiet?

I started to realize that for many of these women—crossing the deserts of the Middle East at a time when a woman couldn't even cross London unaccompanied—travel was a way of breaking free of society's constraints. It's telling that the first female travel writer was a nun. Perhaps it was only by opting out entirely of a traditional familial role that women could escape domestic pressures and devote themselves to other pursuits. By crossing borders they broke away from the narrow expectations laid out before them—wife, mother, spinster. Maybe things haven't changed all that much for women and travel still does that. At least, it feels that way for me.

Back in Tel Aviv airport, with huge posters of a robot-like Kim Kardashian modeling sunglasses bearing down at me, the border guard handed me back my battered passport. I screwed my courage to the handle of my suitcase and headed out into the balmy night air. Tel Aviv was city I'd never been to before. I knew not a soul and could speak not a single word of the language, and obviously I had forgotten to get any local currency. Like I said, I love traveling, but I am very bad at it.

But as I jumped into a taxi and watched the unfamiliar scenery whizz past as an Israeli pop song unfurled from the tinny speakers, I realized I felt better than I had in a long time. Maybe "just me" would be just fine.

CHAPTER ONE
Sister Act

Following in the footsteps of Egeria, the first female travel writer

When I found out about Egeria, I thought it would make a great article, so I pitched it to the editor of a travel magazine I write for called *Suitcase*. I'd told her that I was going to Israel to follow in Egeria's footsteps for a feature, but the truth was, I just wanted to escape London for a little while. I'd emptied the flat Sam and I had shared, and I left it clutching a wok and sobbing in the street. Full-blown weeping in public was nothing new for me anymore. I had been crying so much my eyes were permanently red and raw, and I wondered how I had so much snot in me.

Although at first it had felt comforting to stay settled among the familiarity of our old life, eventually existing among the wreckage got too much to bear. So I had packed everything up, including all the memories of our life together—the wedding cards, the photos of us, the letters he'd sent me. I found a strip of sepia-tinted passport photos we'd taken in Rome; the third one was the obligatory snogging shot. We looked so happy and in love. It was hard to know what to do with these things, hard to work out what had gone wrong.

As a freelance travel journalist, I'm used to going places with little notice and organizing a lot of the itinerary myself. Normally when I'm going on a trip for work, the tourist board of said country bends over backwards to help me find accommodation and flights. But because I'd decided to visit Palestine as well as Israel, the Israeli Tourist Board wasn't too keen on helping me out, which meant I had to organize and pay for everything myself (oh the horror). Although I received expenses from the magazine, they didn't stretch too far, so I planned on a lot of hostels, street food, and public transport. After a long time of being cosseted and swaddled in luxury on press trips, this felt very different, but it was good to be traveling completely on my own terms. True to form, I'd not managed to organize as much as I would have liked to before I left, so I would be very much embracing the risk and chance elements of an adventure. The day before I was due to fly, I read that there had been a rocket strike on Tel Aviv—the first since the 2014 war—so I wondered if that meant I could also add some danger to the mix, too.

Women have many reasons for going traveling alone, apart from seeking adventure. Often we're searching for something, reaching for some meaning in our lives, something bigger and higher than ourselves. We go to ashrams or meditation retreats or yoga holidays, embarking on modern-day spiritual quests that owe much to ancient traditions. It was under these auspices that one of the earliest recorded examples of a woman traveling alone set off into the world.

"I know it has been a rather long business writing down all these places one after the other, and it makes far too much to remember. But it may help you, loving sisters, the better to picture what happened in these places," wrote Egeria, the first female travel writer. I love to picture her, having arrived somewhere in the Levant area between AD 381 and 384, picking up her parchment and deciding to send word back home of what she was experiencing.

Egeria wrote a first-person account of her trip addressed to her *sorores* (sisters), which has led scholars to deduce that she was a nun or some kind of religious figure. However, she could have just as easily been using the word as a term of endearment, addressing a group of Christian women or a troop of female siblings waiting back at home, *Little Women*-style.

Although only one incomplete manuscript preserves her account, known as the *Itinerarium Egeriae* and written many years later in the eleventh century, we still get an idea of this woman's personality and enthusiasm for exploring. The excitement she felt at meeting holy men and women and observing local customs is palpable. We don't know her age, but scholars have deduced from her energy and activities that she was relatively young. She described her "boundless curiosity" and went to great lengths to record the details of what she saw, rarely mentioning the challenges she would have undoubtedly faced.

We know that she was educated, probably comfortably off, and wanted to see the lands of Genesis, the Old Testament, and the Gospels "for the sake of piety." Writing in Latin (and I'm indebted to the various translations of her diary), Egeria stated that she had come "right from the other end of the earth" but didn't say exactly where home was. However, a seventh-century Galician hermit named Valerius (doesn't everyone know one of those?) wrote a letter praising "the most blessed Egeria and her travels," so she might well have come from the same part of Spain. Scholars think she could have been the daughter of a merchant, which would help explain how she came by the contacts (and the cash) she would've needed to organize the tents, camels, mules, and local guides necessary for desert travel in Roman times.

Her journey was self-directed and self-organized, and she delighted in recounting her experiences for her sisters. By her own admission, she was "inquisitive," had a taste for adventure and took hardship in her stride. She climbed Mount Sinai, deep in the Egyptian desert, on foot "straight up, as if scaling a wall."

While the extent of Egeria's letters is unique for a female traveler in this period, she wasn't an anomaly in making such a trip. The period between the fourth and sixth centuries was a high point for religious travel by women. The majority were widows or unmarried virgins who, by virtue of their status or devotion, had more freedom than the average woman of the time. Many traveled with female family members or female friends and gravitated toward other female travelers they met along the way. Egeria formed a special connection with a deaconess she met named Marthana—"surely I cannot write down what her joy and mine could have been?" Not everyone approved though—the fourth-century saint Gregory of Nyssa had this to say about solo female travelers: "It is impractical for a woman to pursue so long a journey unless she has a conductor…on account of her natural weakness…she fails to observe the law of modesty."

Unlike Egeria, who traveled through Constantinople to arrive in Jerusalem, I got a direct flight from London to Tel Aviv (much smoother than a donkey ride). It was nearly 4:00 a.m. by the time I arrived at my towering hotel with hundreds of identical rooms. When I checked in, the receptionist asked me if I was there for work and why.

"I'm a travel journalist," I told her.

"This is very nice job, I think," she smiled.

I relished the reassuring anonymity of staying in a place like this— the strange paper ribbon they put over the toilet seat, the wafer-thin clean white towels, the tiny notepad and branded pen. My balcony overlooked the beach and, despite morning only just breaking, joggers and surfers were out on the sand, and a smattering of disheveled revelers were sitting outside bars smoking. Tel Aviv is known as the Miami of the Middle East, and I could instantly see why.

Although I'd only had a few hours of open-mouth plane sleep (which everyone knows isn't the same as the real kind), I decided to leave the sterility of my hotel room and get out into the sun. I treated myself to lunch at a far fancier hotel than I was staying in and sat outside, eating tiny plates of Israeli salads and *masabacha*, a warm hummus made with white fava beans. I asked the waiter to teach me some Hebrew and, once we'd covered the basics, he told me that if I was going to talk like a Tel Avivian, I'd need to know the word *balagan*, which means "chaos." It felt apt.

I try to always have lunch somewhere special on my first day in a new city. It encapsulates one of my favorite things about traveling solo—that no matter how messy your life is back home, when you're away you can be anyone you want. You can be the kind of person who eats dinner in a five-star hotel and flirts with waiters and then orders a martini at the bar. I could be a heroine in a Fitzgerald novel, surrounded by white-linen napkins and chilled cocktails. Nobody knew that back home, all my clothes were in garbage bags, and while there I wondered if I could treat myself to a takeaway coffee.

I should point out though that I was consciously *not* flirting with this particular waiter—even though like most Israelis I've met he was tall, dark, and handsome. I'd decided that I needed a period of celibacy. Whether it was going to an all-girls' school or too much time spent reading (and writing for) a certain type of magazine, I had always found it hard to shake the feeling that receiving male attention is the ultimate goal. To that end, I'd spent my life leapfrogging from one relationship to the next. I couldn't remember the last time I was genuinely *single* single.

I also felt bruised from what had happened with Guy, the friend I'd started seeing right after Sam, and I had separated. Initially, Guy and I had kept things secret, but as the weeks of heady new relationship bliss turned to, "Oh wow we've actually been doing this for a few months," I realized that I'd started to really fall for him. And then he told me, in an excruciating telephone call, that he'd slept with his ex-girlfriend (who, for bonus pain points, I was also friends with). It felt like he'd put a hand inside my intestines and tied them in a pretty red bow. Oh, and he thought he was still in love with her. Which was great. So although part of me thought that a new love interest might take my mind off things, I wanted to spend at least a few months not being defined by the male gaze, not thinking of myself in relation to a man—not getting laid, basically. So it felt fitting that I was in Israel following the trail of a nun.

After a day in Tel Aviv, I took the bus to Jerusalem and the first thing I noticed was the clothes. Not just the uniforms of the young soldiers (every Israeli does national service when they leave school, so teenagers in khakis is a common sight), which makes it feel disconcertingly like wartime, but also the Hasidic Jews in their mink hats and snazzy gold robes and a group of nuns in white habits. It also happened to be Purim—the Jewish Halloween—so I saw people dressed as clowns, monks, and (the best costume of all) a man dressed up like he was riding a camel, while riding an actual skateboard. I'm pretty sure that's how Egeria would've traveled if she could have.

As I walked around near Mahane Yehuda, the city's central food market, I suddenly got the feeling that I'd stepped back in time. This neighborhood felt different somehow. There was white washing hanging between the houses, a shop selling only black hats, and an almost eerie peace and quiet. I realized I must have been in Mea Shearim, the Orthodox Jewish area. From here, it was not long before the atmosphere got livelier and the colors brighter. Amid the pretty courtyards of Nachlaot was a vintage clothes shop, a few art galleries, and some splashes of street art.

Students from the nearby architecture school spilled out on to the street and, walking among them, I had that pang of feeling exposed and alone and not having any friends—a flashback to the school canteen. I wanted to reach for my phone just to seem like less of a loner but instead I got talking to a young woman named Jenna who was from New York and studying art history in Tel Aviv. She pointed me in the direction of the most-raved-about restaurant in town for dinner, where I sat at the bar and chatted to Tal, the bartender—a beautiful Israeli American with a mane of curls and perfect red lipstick.

Dining alone is one of my favorite things about traveling solo, even though for a long time it was the scariest. Forget sitting at a lonely table with one sad candle and a chair facing the wall—solo dining is the best.

There is no smugger feeling on Earth than rocking up to that hugely popular new restaurant, watching a queue of couples and gaggles of friends getting turned away because all the tables are taken and then waltzing in and snagging the single seat remaining at the bar. And then ordering a really decadent dinner just for one. You also get the satisfaction of not being one of the silent couples sat at tables looking bored and/ or huffy because *she* really wanted to eat at the hotel but *he* insisted on coming here. And you never have to split a pudding.

Tal kept offering me shots of arak—"Let's do a little *l'haim*!"—and telling me how gorgeous I was. In one version of the night, I went home with Tal to her small room above the bar, which was probably full of jazz records and bohemian lampshades. I may have been on a man ban but I'd never said anything about women. But the truth is, either it was the arak or the journey, or my annoyingly conventional heterosexuality, but I was hit with a wave of tiredness, so I made my excuses and went to bed.

The next morning, I forced myself to wake up early and make my way to the Old City. Although it wasn't yet 8:00 a.m., behind the walls life was already pulsating. The Old City has four distinct quarters (Muslim, Jewish, Christian, and Armenian) and many more subsets within. I enter through Herod's Gate and found men smoking *shisha* and playing cards and boys carrying wooden wheelbarrows of *kaek* (a sesame-covered bread). *This must be the Afghani area*, I thought. I bought a loaf and walked down the steep narrow cobbled alleyways, past cafes selling hummus in red plastic bowls, and groups of religious tourists singing hymns on the Via Dolorosa, tracing out the journey that Christ made with his cross two millennia ago, which Egeria followed some three hundred years later.

I joined the queue to place my palm in a smooth hand-shaped part of the rock where Jesus is meant to have stumbled and rested his palm. I liked the idea that Egeria was tracing Christ's route, as I was tracing hers. It also felt like giving a high-five to Jesus. I suddenly had a pang of regret that there was no one to share that (clearly hilarious) thought with. But I guess that's what social media is for.

I carried on past the shouting hawkers and trundling carts to the Small Wailing Wall, a lesser-known, more sheltered part of the Western Wall, where people have crammed folded notes and prayers into the crevices between the stones. As we stood there, two small boys, one in a *kippah* and one in a *keffiyeh*, walked up, hand in hand, and kissed the wall together. It was an almost too-perfect metaphor of hope and friendship amid the ongoing tensions in this part of the world.

Then I reached the Holy Sepulcher—the main church in Jerusalem where pilgrims line up to touch the rock Jesus was crucified on or to go inside his ornate tomb. The strong smell of incense, throngs of people visibly overcome with emotion, and strings of twinkling oil lamps and baubles make it an intense experience, even for an atheist like me.

Watching this outpouring of devotion, it occurred to me that Egeria wasn't really traveling solo. Although she doesn't mention having servants, she does pick up a few guides along the way, and she was constantly meeting monks, nuns, and bishops. She was also, of course, traveling with her faith. She had her god with her for the whole ride. As someone who has never felt drawn to any religion, this realization only made me feel more lonely.

I thought of Egeria again when I spotted the crumbling Roman arches stretching overhead on certain streets in the Old City—the very arches she would've walked under. I passed an alleyway with a sign saying "The Nun's Ascent" in Hebrew, Arabic, and English. The limestone steps were glassy from all the feet that have polished them, maybe Egeria's too.

I somehow navigated my way out of the maze-like Muslim quarter to reach the Mount of Olives, which Egeria helpfully noted is "a very big hill." I passed the rock where Judas was said to have betrayed Jesus, which Egeria also mentioned visiting; there's still a "graceful church" at the bottom, possibly still on the same site, along with the Garden of Gethsemane—now a rather underwhelming patch of olive trees.

Over the three years she stayed in Jerusalem, Egeria made trips to nearby cities such as Jericho, so the next day I got on a bus to the lowest and oldest city on Earth. On the way down through the hills, I spotted Bedouin tribes in corrugated iron shacks and eventually passed under a huge Coca-Cola sign and a stone archway to enter the city, which I knew from my shaky biblical knowledge was the Promised Land.

Archaeologists from Italy were dusting the fortifications at the palace of Tel Al Sultan, which dates back ten thousand years. I thought about hiking to the monastery on the Mount of Temptation where Jesus fasted for forty days and nights, but instead I did the opposite and ate a massive lunch of a falafel *sabich* (stuffed pita) with preserved lemons and eggplant and a fresh pomegranate juice in the peaceful main square.

On the way back, ears popping from traveling 250 meters below sea level, I jumped off the bus at the Dead Sea, which Egeria also visited. I'd been advised by a friend to swerve the charmless resort towns and enter the water at a hidden spot called Ein Kedem instead.

I was a bit hesitant getting in, imagining that the salty water would thrust me upward with force, but instead it was just like floating in any sea, with a bit of added bounce. "Its water is extremely bitter," noted Egeria. "Fish are nowhere to be found in it, and no ships sail there. If anyone goes to swim in it, the water turns them upside down." I rubbed the supposedly magic mud into my skin and noted the one big downside to being on my own. Nobody to take a comedy photo of me reading a newspaper while floating on my back. I made a mental note to buy myself a tripod for future solo travel photo ops.

That night, skin and hair still silky soft from the minerals in my mud bath in the Dead Sea, I took the bus from Jerusalem to Checkpoint 300 to reach Bethlehem. When Egeria visited Jesus's birthplace, it was part of Jerusalem. Now it is part of the Palestinian West Bank, behind a seven-hundred-kilometer dividing wall built by the Israelis in 2002. A few of the Israelis I met in Jerusalem expressed surprise that I was going into Palestine at all, let alone by myself and as a woman. "Just don't do it at night," cautioned one waiter ominously. So obviously it was pitch black by the time I attempted the journey.

I was a bit nervous as I sat at the bus stop. There was no timetable to be found, not that I'd have been able to read it in Hebrew or Arabic, anyway. But eventually the twenty-one arrived and onboard I met Mahmoud, a Christian Palestinian—I noticed that it's normal here for people to tell you their nationality and faith within the first five seconds of meeting you. He sold olive wood sculptures at Jaffa Gate and was making the journey back home.

Mahmoud told me not to worry; he does this journey every day, and he'll help me get through the checkpoint. It was a labyrinthine concrete penning system, which called to my (privileged, Western) mind a mixture of queuing up to get into Glastonbury and going through swimming pool turnstiles. Some parts of the checkpoint were in total darkness, and I had to use the flashlight on my phone to light the way. Though it was pretty much empty and only took me a few minutes, in the morning thousands of Palestinian workers would line up here in order to go to work in Jerusalem, a procedure that can last several hours.

I was excited to be staying at Walled Off, a hotel, museum, and art gallery right next to the wall, which the Bristol-based artist Banksy opened in 2017. Entering the candle-lit, jazz-music-filled lobby after dragging my bags through the narrow turnstiles and dark corridors of the checkpoint was such a jarring feeling that I couldn't help thinking of *Casablanca* or colonial times.

The hotel was surreal and full of Banksy's trademark humor. The check-in desk was labeled "Rejection" not reception, there was a Grecian bust wreathed in fabric made to look like smoke from a tear gas canister, cherubs in gas masks painted on the wall, a lift jammed open with concrete breeze blocks, and a printed note on my pillow from Banksy that said "Welcome to Bethlehem—a place renowned since Biblical times for its inadequate hostelry facilities—a tradition we're likely to continue here."

I was on such a high in the lobby bar, listening to a Palestinian folk singer play a round, flat guitar—a *bouzouki*, he later told me—but out of nowhere I suddenly got a familiar feeling. There was pressure building in my nose like I'd done a forward roll in a swimming pool and, just like that, I was crying again. But it felt different. For the first time, I was crying not for all I'd lost but because of the things I'd gained. Maybe it was the music, or the fantastical setting, or the reminder of how fortunate I am that I can come and go as I please, not only across the border between Israel and Palestine, but all over the world.

That night, I slept more soundly than I had in a long time in my wonderfully chintzy pink room, only stirring with the call to prayer around dawn. After eating an order of shakshuka under Banksy's display of mounted security cameras, I took a walking tour of the wall, adorned with street art. Personal favorites included "Make Hummus Not Walls" and Donald Trump snogging a watchtower.

It wasn't long before we reached Aida Refugee Camp. I had mixed feelings about going there. It felt a lot like "poverty tourism" and yet, deep down, I had to admit that I liked the idea that I was the kind of person who visited refugee camps. For many years, I felt inadequate being a travel journalist. That it was a fluffy subject, and I should be writing about war zones, not infinity pools, that my choice of subject matter was proof that I wasn't clever enough to talk about politics.

Being here in person made the complicated history of the region make a bit more sense. Here's a very brief and extremely simplified recap if you're interested: After the Holocaust and the huge displacement of Jewish people during the Second World War, Britain (which ruled Palestine at the time) gave a large part of Palestine to displaced Jewish people. The Arabs who lived there felt that was unfair, and in 1948 the two sides went to war. In Hebrew this conflict is known as "the war of liberation" and in Arabic it is *Nakba* meaning "catastrophe," which just about sums up the differing viewpoints. When the war ended, an area called Gaza was controlled by Egypt and another area, the West Bank, by Jordan. These two areas contained thousands of Palestinians who had fled what was now the new Jewish home of Israel.

In 1967, there was another war, known as the Six Day War. Israel occupied the Gaza Strip and the West Bank. In 2005, a militant Islamic movement called Hamas won elections and gained control there. Many people call Hamas a terrorist organization because it uses violence to achieve its aims. It refuses to recognize Israel as a country and wants Palestinians to be able to return to their old home. Since then, Israel has held Gaza and the West Bank under a blockade—controlling its borders and limiting who can get in and out.

Being at Aida and seeing how some 6,000 Palestinians have been living in this refugee camp since the war of 1948 brought the political situation to life. It felt like a permanent village, not a temporary shelter. I met Layla, who was selling bars of olive oil soap and jewelry made from the shells of tear gas canisters. There was more street art and a sobering list of children killed "during an Israeli massacre in 2014" painted on a black wall.

Before I left, there was just time to walk down Star Street, through Manger Square, to reach the Church of Nativity. As I watched pilgrims gather for mass, in the spot where Jesus is said to have been born—just as they would have congregated when Egeria was here—I was reminded that both our journeys were inspired by writing. The Bible was Egeria's inspiration; she was following in the footsteps of Moses and Jesus, just as I was stepping into her shoes. By documenting her travels, this fourth-century pioneer hoped to conjure up her experiences in the minds of her "dear sisters" at home. Maybe this is the power of writing—that it can become a form of telepathy, as well as a way of traveling through time. I hope that I can do something similar with this book. *Inshallah*, as they say around here.

TIPS FOR SOLO TRAVEL FIRST-TIMERS

Traveling alone can feel daunting and, even when you're used to it, the first few hours/days can be tough. There's no one to make conversation with, no one you need to think about. Suddenly it's just you and your thoughts. It feels empty and a bit lonely at first, but you soon adjust and start to enjoy the pleasure of taking yourself on holiday. A 2015 study from Oxford University found that people consistently underestimate how much they enjoy their own company. If you haven't been on a trip alone before, here are some ideas that might help ease you in.

1. **Go with an organized group.**
   I know, even the idea might make you shudder and think of troops of people filing off a coach bus and trudging behind someone holding a flag. But there are now some really cool tour companies that cater just for solo travelers of a similar age—G Adventures and Flash Pack, for example. A few, such as Thelma & Louise, are solely for solo female travelers. Look for an itinerary that gives you flexibility and free time when you want it, rather than one where the whole day and evening seems packed

with trips and activities. Going with a group is an accessible way in for a first-timer, particularly for more difficult places to navigate alone, such as China or Iran.

2.   Book an activity holiday.

Learning to do something with other people you don't know is a great way to get a taste of solo travel while still feeling like you have company. You'll be able to retreat on your own when you want solitude, but there'll still be people around to chat with at meal times. It doesn't matter what the activity is—painting, cooking, playing tennis, surfing—or how good you are at said activity. That's the beauty of learning it with a load of randoms who you'll probably never see again.

3.   Pick your destination wisely.

As someone who once spent ten days solo on a honeymoon retreat in the Maldives, watching couples smooch on sun loungers and take selfies in "Just Married" bikini bottoms, I can tell you that it's often a good idea to stay somewhere designed for people with no plus one. Japan is famously ideal for the solo traveler—there are plenty of hotel rooms and restaurant booths designed for one person and things to do 24/7. Safari can be tricky by yourself—the cost alone makes it almost exclusively for "the newlyweds and the nearly-deads," as one safari lodge manager in Botswana put it to me. Also, being in a tent and listening to lions roar outside is even scarier when you're alone.

4.   Choose your accommodation well.

Whatever your budget happens to be, luxury hotels—where staff are trained to give you maximum privacy and leave you alone—or little boutique hotels, where you're the only guest, probably aren't the best option for a first-time solo trip. Hostels (more on these later) are a great way to meet people because they have so many events and communal areas. Or consider a co-living or co-working space that has shared office space and kitchens alongside your own apartment. Outsite and Norn are just two examples of this growing trend.

5. Book something for your first night.

I'm definitely a go-with-the-flow type traveler and believe that wonderful things happen when you leave it up to chance. But I've come to realize that, even if you're an experienced jet-setter, it's good to have a plan on your first night in a new place. Spontaneity plus jet lag plus being alone can equal wandering around in the dark for hours until you're forced to eat in the Hanoi branch of KFC because there's nothing else open and then bedding down in the first dump you happen upon because you're too tired to look for anywhere else, and it's too gloomy to see the dirt. Start your trip off right and know where you're staying on the first night and maybe also where you'll be eating. Just those two things, if nothing else. It makes a massive difference.

6. Think carefully before committing to a travel buddy.

It's very common for solo newbies to latch on to the first other solo traveler they meet and then make a plan to travel together for the rest of their trip. This can be wonderful if you're craving company and, who knows, they might well be your soul mate. But don't commit too soon. You might regret boarding a sixteen-hour coach trip or booking a week's accommodation with someone who you soon discover has an annoying laugh and/or won't stop talking and/or reminds you of someone you hate back home. Being with the wrong person feels much lonelier than being on your own. This is true of relationships and not just traveling, by the way.

7.   Don't beat yourself up if things don't go as planned.

We all have the best of intentions to find the raved-about secret taco truck or that hole-in-the-wall cocktail bar only locals know about, but some nights you will end up in the Häagen-Dazs café, and that's just fine. Don't feel like you have to tick off every must-see sight in the guidebook on your first or last day—or at all. Once in Bogotá, with seven hours to kill before a flight and the best of intentions to hike up Cerro de Monserrate and go on a street art tour, I spent the whole time in a coffee shop using their free Wi-Fi, and I don't regret a single second of it.

8.   Think about your inner voice.

This is a good tip for life, not just solo traveling. Pay attention to that inner dialogue we all have. Is it supportive and comforting? I think women especially tend to be our own worst critics, talking ourselves down and talking ourselves out of good ideas. As Maya Angelou put it, "I learned a long time ago that the wisest thing I can do is be on my own side." I didn't learn that a long time ago, but I wish I had. When you're alone, your inner voice might be all you've got to see you through. So make sure yours is saying good things. If it's not, you can consciously change it until it does. You will feel silly doing this at first but just try a few "You're doing so well" type words of encouragement that you'd give to a friend in your situation.

9.   Train yourself up at home.

Being comfortable in your own company is like a muscle, you have to use it or lose it, so practice spending time by yourself. Go on small solo adventures and see how you find it. Take yourself for a coffee alone. Visit an exhibition all by yourself. Then eat a meal out. Then eat that meal in a proper restaurant. Then eat that meal in a proper restaurant without looking at your phone for the entire time. Okay, now you're ready.

CHAPTER TWO
Sleepless in Shanghai

On the trail of monkey-bothering badass Emily Hahn

A few weeks later, I arrived in China with glitter in my hair, face paint on my neck, and false eyelashes stuck somewhere about my person, having had zero sleep after a friend's drag-themed fortieth birthday party in Cornwall. Getting from deepest darkest Penzance to Shanghai was a ludicrously epic mission. Having not been to bed at all, it felt like a waking dream of gate numbers and terminal letters; I just had to transport my body to the correct place at the correct time in order to get where I needed to be. I wouldn't recommend it. And my thoughts were somewhat in transit, too.

I had been going through a strange phase in which I woke in the middle of the night and replayed random moments with Sam. I thought about how his breathing sounded when he was asleep. About his heartbreakingly generous acts of kindness. About odd fights we'd had in otherwise forgotten restaurants. It was strange to be having all these Sam-related thoughts now, months after the end of things. But maybe big events are too much to process all at once, and so this was the slow drip-feed of sadness that my body and mind could cope with.

It felt weirdly appropriate that I was arriving in China feeling like this, as, the last time I was here, four years ago, I was similarly discombobulated. Although everything back home seemed to be going brilliantly—my freelance career was taking off, we were newly engaged and had just bought a flat together—this particular trip was especially chaotic, even by my standards.

Having not realized that you needed a visa to go to China (I told you I was bad at this traveling lark), I ended up having to meet the rest of the journalists on my trip three days late. "It's fiiiine," I reasoned to myself (and my slightly worried editor when I called her from the Chinese embassy). "I'll only miss the Great Wall of China part and that's just, like, a wall."

When I arrived in Beijing, I was collected by a driver who spoke zero English, nervously clutching a sign with my name on it. When we were in the car, he got Ricky, the Chinese guide who was leading our trip, to speak to me on his phone. Ricky (his Western name) told me to meet him and the rest of the group at a restaurant called Cha Fing at Beijing train station. "You can't miss it," he told me.

Nearly two hours of crawling along smog-choked highways later, we pulled up at what I guessed was the train station, and my driver politely indicated for me to get out the car by dumping my rucksack on the pavement. I surveyed the scene. There were so many people—Beijing's central station at rush hour—but no Cha Fing. Nothing in English at all, actually. For some unknown and very inconvenient reason, everything was in Chinese.

My phone didn't work here. I'd already made a few feeble attempts to do important things like look at Facebook, which refused to connect. I found something that looked like it might be a hotel. I made my way through the lobby and asked the receptionist to borrow their phone. After interminable minutes of them looking at me incredulously, I did the universal mime of "the phone hand" until they acquiesced.

I dialed the contact number I'd been given for Ricky with trembling fingers. I was greeted by an irritatingly cheerful Chinese operator, presumably telling me to go fuck myself. I walked back out in the square, trying to stay calm, and flagged down two passing police officers—surely they would be obliged to help me? But after a few minutes of me saying "Where's Cha Fing?" and them looking baffled, we all realized we had no way of communicating with each other so they shrugged and walked off, leaving the crazed woman with the huge rucksack to speak gibberish to someone else.

At this point, I realized I could be in serious trouble. It was getting dark. I had no food or water. I knew that the train we were supposed to get to take us to the next city on our itinerary left in less than an hour. To start with, I couldn't even get inside the station because it turned out that you needed a ticket to get in, but I looked so desperate that I somehow managed to convince the guard to let me through without one. Perhaps the fabled Cha Fing would be *inside* the station? I went up and down all six escalators—but there was no sign of it. No sign of anything even remotely recognizable or intelligible or helpful to my cause.

I realized the immensely privileged state I'd traveled in throughout my whole life, where I'd taken for granted that I could go anywhere and be understood. All the places I'd been where people had made the effort to speak my language. How much of my traveling life was dependent on Google Maps and Google Translate and other things that were banned in China.

I saw a group of teenage girls and thought maybe one of them might speak a little English or had listened to enough Justin Bieber that she might understand me enough to lend me her phone so I could try to call Ricky again, or at least someone in London. "Please help me!" I begged them. They put their hands up to their mouths and giggled nervously. This wasn't going well. "Where. Is. Cha. FING?" I did that embarrassing thing all English people do when they're not being understood in a foreign language so they speak very loudly and annunciate every single word.

As an extra livener, I had absolutely no money to my name—the currency I'd got in a hurry as I zoomed through the airport turned out to be Japanese yen, not Chinese yuan. (They sound irritatingly similar to people who don't speak Japanese or Chinese, don't they?) I discovered this when I tried to buy a bottle of water at Beijing airport and the man held up my banknote and laughed. Perhaps I've just given him a really high denomination, I thought, like the typical tourist who tries to buy a flat white with a fifty pound note.

"No problem," I reassured him, offering him a different sheet of paper with another stern-looking man and characters my English-speaking self couldn't understand on it. The cashier laughed again and then looked at me with pity before calling his friend over to look and laugh. "I can see my money is no good here," I said, snatching back my cash, like I was Julia Roberts in *Pretty Woman*, if she happened to be in an airport kiosk selling overpriced refreshments not Rodeo Drive.

I tried to use a cashpoint at the train station but it spat back my pathetic non-Chinese card in disgust. I had no legal tender and no way of contacting anyone. I had to accept that my situation was pretty hopeless and perhaps I should go back to the airport where maybe one person might speak enough English to let me book a flight back to London on my credit card. I was wondering how exactly I'd get a taxi with no money when I saw something.

It was a huge flock of white people sauntering out of one of the restaurants flanking the station. In fact, the exact place where I'd been dropped off. They stood out a fair bit, what with the fact they were a foot taller (and also wider) than everyone else and, well, white. I get why Chinese people call white people ghost face. I certainly looked like I'd seen one.

"Oh Miss Kate, there you are!" said Ricky, our guide, nonchalantly. "We were waiting for you in Cha Fing…" He gestured to the squiggly black lettering on the sign on the restaurant's facade. "Oh, that's funny I always thought that sign was in English," he chortled. As if this was a minor and amusing inconvenience and not the cause of the most stressful hour of my entire life.

I wiped my tear-strewn, haggard face and had to shake hands with all the other journalists and pretend to make light of what I'd just been through. I then proceeded to borrow money from all of them. Welcome to China!

I'm glad that I went on that disastrous trip, though, because it was while in Shanghai back in 2016 that I first heard about Emily Hahn, the woman whose journey I'd come back to retrace three years later. Ricky had taken us in the pouring rain to the former offices of the *China Daily News*. One of many imposing art deco buildings on the Bund, it has frescoes dedicated to art, science, commerce, truth, and journalism.

I noticed that the last of these frescoes seemed to depict a man writing on a tablet surrounded by half-naked women looking on in admiration. This was actually quite close to the fantasy of a lot of the male editors I'd encountered on the national newspapers I'd worked for. Ricky told us about the building, now the office of an insurance company, and some of the notable people who had worked there in the 1930s. Under the cover of his umbrella, he lifted up his iPad and showed us a picture of the coolest person I'd ever seen.

It was a black-and-white shot of a woman with a short, soot-black crop; pale skin; and pouty lips. I thought she looked like a silent film star, one of those 1940s femme fatales. Oh, and of course, she had a pet monkey on her shoulder. When I found out she was a travel journalist who'd written more than fifty books and had a life more interesting than any soap opera, it seemed like a tragedy verging on farce that I'd never heard her name before.

When I got home, I couldn't stop thinking about Emily Hahn. *New Yorker* magazine, the publication she wrote for in a career spanning eight decades, described her as a "forgotten American literary treasure." I started reading her memoir *No Hurry to Get Home*. Well, it was supposed to be her memoir but, as she explains in the prologue, she couldn't be bothered to write anything new so just repurposed some of her best *New Yorker* essays. Here was a girl after my own heart. Will work, would rather not. I couldn't love her more if I tried.

The essays are a rollicking romp through the twentieth century. Known as "Mickey" to everyone, Emily Hahn wrote of being the only woman on her engineering course at university and of attempting suicide while living alone in New York during the Great Depression. In 1930, she left the United States for the Belgian Congo because, "I was young and impulsive, because I'd always wanted to." She hitchhiked across central Africa alone. She moved to Shanghai and became the concubine of a Chinese poet. One essay in her memoirs starts with the inimitable line: "Though I had always wanted to be an opium addict, I can't claim that as the reason I went to China."

My own reasons for coming to China for a second time were partly to get to know Emily Hahn a bit better and partly to distract myself from my troubles. As well as my Sam memories bubbling up, I was also still dealing with the fall-out from what had happened with Guy—a person I'd always been attracted to but who seemed wildly unsuitable in almost every respect. With curly black Jim Morrison hair and a love of hard drinking and even harder partying, he had seemed like a classic rebound from the safety of my former life with Sam. He even had a motorcycle. My friend Sarah described him as "the hunk in a teen movie." You get the idea.

Since the whole sleeping-with-his-ex thing, Guy had come clean to her about our illicit relationship. So now she was emailing me, understandably pissed off, and lots of our mutual friends were avoiding me. I'd tried to tell myself that I hadn't been *that* good friends with Guy's ex-girlfriend—and they were definitely broken up—but it still felt like an extra dose of indelible shame, when I was already pretty full of self-loathing after bolting from my marriage.

I felt like I was dealing with two break-ups at once. Even though everyone told me that I should just "enjoy being single!!!" and "make the most of this time," they didn't know the fear and loneliness of staring down the barrel of another whole weekend with no plans. When you're in a relationship, you don't have to make plans. You wake up late and have brunch and shop for food together. Then you might decide to see a movie or go out for dinner or get takeout. But when you're single and you have no plans, life can feel desolate. Without a nine-to-five job to ground me or give me any routine, I'd spend hours alone on my laptop and then realize quite suddenly that it had got dark. I tried hard to relish all this newfound time I had, but it felt impossible to enjoy it when all around me friends were announcing pregnancies and engagements and house purchases.

What made the whole situation doubly depressing was that for all the years I'd been with Sam, I had felt so ahead of the game. While my single friends cried on my shoulder in their twenties, wondering if they'd ever find someone, I was all smug with my stable relationship, my beautiful flat, my great job. But now here I was. If life was a game of snakes and ladders, I had slithered all the way back to square one.

At my lowest moments it felt like days stretched on, and I just had to get through them, but for what? I believed I'd made a massive mess of everything. I'd been given so many opportunities and privileges and had done nothing good with any of them. There was one particularly dark moment where I calmly and decisively reasoned with myself that ending my life wasn't an option, so all I could do was keep going. Everyone told me that time is a great healer. That "this too shall pass." And I knew it was true. But how long would that take, exactly? Because this felt unbearable.

Now that I was in China again, I tried to focus on Emily Hahn and to make myself feel better by reading about her similarly chaotic life, dating playboys, property tycoons, and poets, shamelessly shocking Shanghai society. But in my current state, life didn't feel glamorous and exciting—more lonely and pathetic. So much had happened since the last time I was in Shanghai, and being back there only highlighted all the upheaval I'd been through. On my first visit to China, I had just got engaged and we were excitedly furnishing our first home. Now I was getting divorced, and home felt like a series of hotel rooms.

After Guy's bombshell phone call, I had blocked his number. He had persistently written me letters (actual letters, maybe he *was* from a teen movie?) and bombarded me with emails until I told him firmly to give me a few months' space. I couldn't decide if I'd had a lucky escape from a bad boy who I shouldn't have got involved with in the first place or if I'd just lost the next great love of my life. I suspected the reality was somewhere in between. I needed some distance from it all, both literal and metaphorical. And the great thing about travel is it focuses you in the moment. As Emily Hahn put it, "The mind of a traveler has only one spotlight and it is always trained on the present scene."

Aside from the Cornwall to China odyssey, my entry into Shanghai was notable for another reason. Although I had managed to get a visa this time, I'd had to sell my soul to get free flights here. I was in China to review the maiden voyage of a cruise. I'd never been on a cruise before, but I had pretended to have read enough David Foster Wallace to know that this was probably going to be funny and terrible in equal measure. Yet nothing could've prepared me for what unfolded.

As the car pulled up at the boat, which was the size of several buildings, the publicist responsible for getting publicity for the ship breathed, "Isn't she beautiful?"

This was *her* inaugural sailing, and for this reason the boat was full of Chinese journalists, and we weren't actually going anywhere. We were just leaving Shanghai, sailing around China for a few days, and coming back again. But that was okay because there were four theatres, a climbing wall, a running track, a surf simulator, a robot bartender, and a skydiving experience on board. When I had decided to take the job, I'd figured that if the worst came to the worst, I could at least just get drunk on the free booze by the pool, not realizing that the pool had a live steel drum band playing Ed Sheeran covers at all times of the day and night.

There was a woman paid to stand by the entrance to the buffet and say, "Washy washy!" to encourage you to wash your hands, which would come to take on a sinister echo in the months to come. At the buffet, I realized that it is possible to be offered more types of food than I could ever imagine, presented in ways I didn't realize were possible—a boat made out of parmesan! A watermelon carved into a turtle!—and yet not want to eat a single mouthful of it, because weirdly it all tasted the same. It tasted like cruise.

I had pictured a luxurious suite but was instead cooped up in a tiny cabin down an endlessly long carpeted corridor. I had an "ocean view" but looking out the window and seeing nothing but blue on blue for days, was, it turns out, the stuff that fever dreams are made of. Yet it paid for my trip to China, and I was very lucky to be there. Still, when we docked back in Shanghai, I could have kissed the ground.

I wondered if Emily Hahn felt similar when she arrived in the city in 1935. She lived in what was then the heart of the red light district on the Jiangsu Road, so once I'd disembarked from my floating prison, I headed there straight away. Emily loved this "crowded, screaming street," which during her time in the city was not just synonymous with brothels but with the fashionable set of poets, artists, and thinkers she hung out with.

Emily's trip to China was meant to be no more than a holiday en route to West Africa. Instead, she would live in Shanghai for five years during the roaring twenties and become a celebrated and eccentric fixture on the city's famously decadent social circuit, dining with playboy millionaire hoteliers like Sir Victor Sassoon and appearing on the society pages accompanied by her pet gibbon, Mr. Mills, whom she would bring to the table suited up in a nappy and miniature dinner jacket. She had a love affair with the Chinese poet Sinmay Zau who gave Emily her first puff of opium.

Jiangsu Road still felt surprisingly shabby, considering how close to the Bund's luxury hotels and skyscrapers it is. Amid a collection of tangled telephone wires and exterior air conditioners, I saw a green door with bicycles outside. The door was open, so I peered inside into the gloom. During the Cultural Revolution, which began in 1966 and lasted a decade, every office, mansion, and building in China was made into communal living spaces, so in Shanghai it's still not uncommon for whole families to live in one room and share toilets and use makeshift kitchens in corridors. In a mansion block in the French concession on my last trip, I had seen a toilet with nine individual flushes and a sink with nine taps, so that each family would use only their own water supply.

I was pretty shocked to notice that in this building they still used metal bedpans, like you'd find in a hospital, and that there was a big trough in which to empty them because they had no indoor plumbing. It summed up the dichotomy between old and new in Shanghai—a place where men riding bicycles piled high with chests of drawers and chickens cycle past huge video billboards playing the latest Gucci campaign. In one of the most technologically advanced cities in the world, many inhabitants still live as they did centuries ago.

From Emily's street, I walked the short commute to her office at the *China Daily News*. On the Bund, all the monolithic marble buildings make you feel like you're in Gotham City. I passed the Fairmont Peace Hotel, once owned by Victor Sassoon, who photographed Emily naked and showered her with gifts. With its curlicue iron doorways, stained glass ceilings, and marble mosaic floors, it's one of the few art deco gems here that hasn't been turned into a bank or the office of an insurance company. This strip is also a popular site for wedding photo shoots, and I counted five of them happening as I walked along, the bride and groom sweaty in huge meringue gowns and full suits in the eighty-five-degree heat.

Just a few blocks away from the gleaming skyscrapers and sports cars of the Bund, I reached labyrinthine alleyways. This was the old town, where geriatric men in their underpants leaned in doorways eating dumplings. An actual cobbler was making shoes by hand at a chair and table. Washing and tangled electrical wires were strung between the low-slung buildings. Last time I was here, Ricky told me that these people were being offered huge sums to move out of their prime real estate and that, before long, these parts of the old town would have all disappeared. It's good to see that some of the inhabitants are still clinging on.

I jumped on the immaculate metro to the former French concession, where the light, airy cafes and plane trees make this small part of China feel more like Paris. It's incredible to think that up until 1842, Shanghai was a small fishing village. After the Chinese were defeated by the French and the British in the Second Opium War, it became a city carved up into autonomous colonial concessions run by the French, the British, and the Americans. Only the walled old town, which I'd just left, remained Chinese. Shanghai became an important cosmopolitan center, the "Paris of the East," famous for having the best art, architecture, dance halls, brothels, clubs, restaurants, and a racetrack.

In 1968, Emily Hahn wrote *The Cooking of China*, a recipe book of Chinese food (including a chapter on "Gentle Teas and Strong Spirits"), so I decided to take a food tour of Shanghai and sample some of her favorite dishes. Over the course of the evening, I learned the art of not slurping a soup dumpling (you nibble off a tiny bit to let the steam escape and then down it in one). I ate a pepper so spicy it turned my mouth numb and discovered that, before anesthetic, dentists used this method instead. I tried jellyfish and algae and washed them down with *baijiu*—at 50-percent proof, it's the strongest spirit in the world. I found out that food is such a pivotal part of life here that a common greeting translates as "Have you eaten rice yet?" I also ate a rabbit's head that night—eyeballs, brain, everything. Once I'd gotten over the jelly-like texture, it tasted nourishing and sweet.

It occurred to me later that when we tell ourselves we're a certain type of person—in my case the "I eat ker-razy food! I'm up for anything, me!" type—it becomes a self-fulfilling part of our character. Something not inherent, necessarily, but created and willed into being. Perhaps relationships are like that too. We convince ourselves that this random other is our person, our soulmate, our other half, and in believing it, eventually it becomes so. Or just doesn't, in my case.

After all that eating, all I want is a long bath rather than the Long Bar, but I'd promised myself I'd visit this famous Shanghai landmark, which Emily Hahn retreated to after being caught up in a Japanese bombing offensive in 1937. Made out of raw mahogany, this thirty-nine-foot-long bar stretches through what was formerly the Shanghai Club, opened in 1900. It's now a Waldorf Astoria hotel and on the night I was there, it was mainly middle-aged men in khaki shorts and T-shirts and Chinese couples gazing at their phones. Quite a contrast to when Emily was there when "it hummed with discussion and rang with cries for whiskey and yet more whiskey." It still retains a bit of *fin de siècle* glamour though, thanks to waiters in tuxedos, filigree lamps, a mirrored oyster bar, and a live jazz band.

I took a seat at the waterfront end of the bar, which, according to the information printed on the menu is where the big bosses would sit, while newcomers to Shanghai—known as "griffins"—would be relegated to the murky depths of the dark wood-paneled room. Along with all the opium, Emily Hahn's memoirs are filled with endless champagne and martinis. So I order a dirty one in her honor.

Emily kicked the opium habit and moved to Hong Kong before the outbreak of the Second World War. The party was over in Shanghai. The Japanese occupied the city in November 1937, foreigners fled, and a civil war broke out. Communism took over and Shanghai was closed off to the outside world for thirty years, enduring famine, drought, and repression. It wasn't until the 1990s that the city reinvented itself again to become the driving force of China's economy—the "head" of the Chinese dragon.

While living in Hong Kong, Emily had an affair with Charles Boxer, the chief spy of the British Secret Service. According to a *Time* magazine article, Emily "decided she needed the steadying influence of a baby, but doubted she could have one. 'Nonsense!' said the unhappily married Major Boxer, 'I'll let you have one!'" Their daughter Carola Militia Boxer was born in 1941.

Shortly afterward, Boxer was captured by the Japanese, who had taken Hong Kong in December 1941, and held in a prisoner-of-war camp. When Emily was brought in for questioning, she was asked by Japanese guards why she'd had a baby with Boxer. "Because I'm a bad girl," she reportedly quipped. When Boxer was eventually released, they married in 1945 and moved to Dorset, England, and had another daughter, Amanda. Domestic country life, however, was not for Emily Hahn. In 1950, she moved to Manhattan, got a job at the *New Yorker*, and created a life that combined nine months in New York or traveling on assignment with three months in England with her family. If that's not the ideal work/life balance I don't know what is.

Emily Hahn lived to ninety-two and wrote fifty-two books, as well as hundreds of articles and short stories. "She moved from here to there and everywhere, like some kind of beautiful, multicolored and quixotic butterfly," wrote her biographer Ken Cuthbertson, "and travel was her nectar." Although I love the idea of being a literary butterfly, at the moment I was still very much in my chrysalis phase. But I had written down an Emily Hahn quote that felt important. In her 1944 book *China to Me*, Hahn wrote: "I have deliberately chosen the uncertain path whenever I had the choice."

I think that's what I saw in her, all those years ago, on an iPad in a damp doorway on the Bund: an enigma full of daring and devil-may-care attitude. And now, while all around me friends were settling down and starting their families, I felt like I was embarking on an altogether more uncertain track. Sometimes the path didn't just feel uncertain. Sometimes it felt like there wasn't even a path there at all.

But reading Emily's writing and getting a sniff of her life in Shanghai made me realize that there was more than one way. I was only just starting to unravel the untold stories of so many unconventional women who had traveled the world and lived life on their own terms. It gave me the confidence to think I could forge my own path, too. I knew that I had always been seeking something just beyond who I was and what I had and what was expected of me. Something beyond baby showers and loft extensions and weekends in Ikea. Now I had the opportunity to explore it.

HOW TO SURVIVE THE NIGHTS WHEN YOU TRAVEL SOLO

Once I got used to being a lone ranger, I generally found that daytimes were easy to navigate. There are sights to see and brunches to eat and sun to bathe in. It's when said sun goes down and you're alone that it's easy to feel lonely and so just head back to your hotel room and eat Pringles while you watch BBC World on repeat. But although a bit of downtime eating extortionately priced minibar snacks is great, you miss out on a lot in a city if you don't enjoy the night times, too. Wherever you are, there are usually ways to find solo evening entertainment that doesn't involve being sandwiched between canoodling couples watching the sunset or propping up the hotel bar alone downing whisky and weeping into a bowl of nuts.

1. **Choose your evening meal wisely.**
Don't pick the fanciest restaurant in town where it's all pristine tablecloths, silver cutlery, and couples over seventy. Look for communal eating experiences where you'll meet locals and other travelers. Food trucks, markets, hawker stalls, and any kind of street food is a good idea (depending on where you are, remember to only eat in places that look busy, and if you're worried

about food hygiene I generally find going vegetarian is the safest option). Investigate whether there are any supper clubs happening in town or restaurants with communal tables. Breaking bread with people is scientifically proven to improve your mood, and nudging elbows over delicious bowls of something slurpable is a surefire bonding experience.

2.   Spend a night at the museum.

Lots of museums and galleries have late-night openings at least once a month, which are typically less busy and have fewer screaming children than during the day. Often they serve wine. There's something that feels naughty and forbidden about being in a gallery at night. Maybe because you're looking at a Picasso while getting quietly pissed.

3.   Check out the gig economy.

Check out local listings and go and see some live music. Music is much better than theatre or cinema because there tends to be no language barrier. Even if the band turns out to be terrible, it'll probably still be a funny experience slash opportunity to people watch. Sofar Sounds is a brilliant movement that hosts live music in intimate spaces around the world. Or maybe make like Emily Hahn and go to a little jazz club where you can sit in a booth in the dark and feel very glamorous. Don't rule out seeing some opera or ballet or classical music. When I was in Puerto Rico, the eccentric, eighty-something, parrot-keeping owner of my hotel very kindly gave me her ticket to see the Puerto Rican Symphony Orchestra. I nearly didn't go because I think I was still hungover from a visit to a rum factory. But I'm so glad I did. The concert hall was spectacular, the music was deeply moving, and I sat next to a sweet American couple, and we all went out for empanadas afterward. I rarely ever go and watch classical music at home but when you're away it can be a surprisingly special experience.

4.   Take a night tour.

Finding a nocturnal group activity is an easy way to fill the evening, and lots of city tours take place at night. Whether it's a nighttime street food tour of Austin or a ghost-hunting trip around Nottingham, they provide a new perspective and a safe and easy way explore a city after dark.

5.   Make the most of technology.

There are now a host of apps that connect you to like-minded solo travelers (Solo Traveler) or locals who'll cook you dinner (Mamaz Social Food) or just cool events and other people who want to do them (MeetUp). If you're in a big city, you're never far away from fun. I know one married travel journalist who even likes to get on Tinder while she's traveling just to meet locals and have a night out (or, at least, that's her story and she's sticking to it). Instead of staring at your phone and disconnecting with what's around you, make technology work for you by letting it link you up with people IRL.

6.   Go out-out.

Yes, I know. Sounds terrifying, doesn't it? Going clubbing on your own? Is she mad? But hear me out on this one. It's actually a really cool thing to do. In fact, you're apparently more likely to get past the doorman at Berghain—Berlin's famously hard-to-get-into nightclub—if you're on your own. If you can survive the initial discomfort, just heading out to a bar or club and seeing where the night takes you can be quite the experience. I guarantee you will meet people—whether it's in the toilet queue, at the bar, on the dance floor. But maybe you don't want to socialize, and you just want to get dressed up, go out, dance, and get really sweaty. Club entry is often cheaper than a high-end gym class—look at it that way.

CHAPTER THREE
On Assignment

Making headlines with "globe-girdler" Nellie Bly

"So you basically get paid for going on holiday then?" is the response I'm met with a lot when I tell people I'm a travel journalist. I tend to laugh and bat them off by talking about the hard parts. How I'm not just lazing on the beach all day but intrepidly tracking down a story, be it the best bars or the black rhino. But the truth is, my job is amazing. I don't know what I did in a past life—maybe saving a truckload of kittens or nuns or something—because when it comes to the job lottery, I hit the jackpot. Which feels like a fitting analogy because most of the places I get to stay in for work make me feel like I'm doing an unconvincing job of impersonating a millionaire.

I've lost count of the number of palatial hotel lobbies where I've sheepishly slung my battered rucksack at the feet of a surprised porter who is clearly more used to lugging Louis Vuitton trunks. I've felt like I'm a *Catch Me If You Can*–style imposter, sipping champagne in a hotel suite that costs more per night than my whole year's rent for my London flat (this was the Bulgari in Knightsbridge, where my suite would've set me back a cool £10,000, and the bathroom was so big it had its own steam room). Yes, there are challenging parts but, for the most part, getting paid to travel the world and then write about it feels like just about the jammiest job there is.

It is a strange job, though. And not just because you can be buying a frozen meal for one on a drizzly London evening when you get a phone call from an editor asking if you can go to Iceland tomorrow and for a split second your mind goes to the supermarket and not the country. Traveling alone to new places for work is exhilarating and tends to be my favorite type of assignment. But occasionally I'll go on what's known as "the Group Press Trip." This is where a number of travel journalists assemble to review a destination or a hotel, usually accompanied by the publicist responsible for promoting it.

I don't know of many other occasions where around ten adults— usually all complete strangers—have to go on a holiday together. Except maybe *Love Island*. And, added to that, the holiday is also work but one person is responsible for making sure everyone has THE! BEST! TIME! It's kind of like a school trip with copious amounts of free booze and never staying in one place for more than two nights. One journalist I know calls them "luxury hostage situations," and I've been on some surreal ones in my time.

Firstly, you will feel like Prince Charles when your host country insists on putting on their national dance for you (which once took over three hours and occurred at the airport right after I'd got off a fourteen-hour flight), or when you're encouraged to meet a local tribe and shake their hands one by one, while trying not to look at whether they've got anything on underneath that leather belt. "And what do *you* do?"

Then there's the most dreaded part of any press trip: "the hotel tour" where you must learn to feign interest in even the most drab and generic "business center" as the over-enthusiastic hotel manager insists on showing you every type of room in their 365-bed establishment despite your chronic jet lag. It's no wonder that on press trips, so many journalists develop sudden migraines, food poisoning, and urgent phone interviews they have to do that minute.

Often the itinerary will be incredible, exactly the kinds of things you would want to see and write about in said location, with enough downtime to enjoy them. Other times you think the publicist must have been stoned and/or sadistic when they organized an 8:00 a.m. tour of a tile factory. Many is the time I've donned a hair net and blue plastic shoe coverings to visit factories processing mango, salmon, guitars, and chocolate. (OK, that last one wasn't so bad.) Distilleries and vineyards are another press trip classic. For some reason, I have little memory of the trip when I toured the Bacardi rum factory in Puerto Rico.

More often, the itineraries are micromanaged down to the last second and are so whistle-stop that they make you want to weep: "Here's an example of a beautiful beach where you can swim with wild dolphins. If you could be back on the coach in ten minutes please." One journalist I know went on a trip to South Korea where the itinerary was so micromanaged that the publicist made them literally sprint through a museum so that they could tick it off the list.

The group dynamics can be pretty interesting to behold. Especially the culture clash brought about by "international" trips that bring together journalists from many different countries. You may or may not be the only English speaker. I've learned that sometimes it's preferable not to be able to communicate with your fellow attendees, after the Japanese journalist I sat next to at a dinner in Paris wanted to use his limited English to talk about The Beatles for hours on end. I've seen an American and a Canadian almost come to blows over dinner because one was "looking at the other funny."

Often the location is incredibly beautiful and the activities bonding, so it's not surprising that press trip hookup are commonplace. On one trip to Ibiza, I witnessed the kind of performative snogging I've not seen since high school between the hotel publicist and one (married) journalist. On a group press trip to Massachusetts, all the other journalists dropped out last minute (lost passports, family illness, a fall down the stairs), leaving just me and a male radio journalist who had split up with his long-term girlfriend the night before. At every luxury restaurant, hotel and, yes, guitar factory, people asked if we were on our honeymoon or presumed we were a couple. Which would've been awkward enough even if he wasn't detailing his heartbreak in minute detail to me during our "romantic" candlelit dinners for two.

Sometimes you wonder whether journalists have been sent on these assignments by their editors as a punishment, not a perk. I went on one trip to Colombia where the Australian member of our party brought along all her own food—bags of nuts, cereal bars—and proceeded to unwrap them at the dinner table and eat them. She couldn't possibly consume the food here because she'd heard bad things about the hygiene in foreign countries. I went on a sailing trip around the Ionian Islands with a journalist who claimed to be allergic to everything, including sunlight, and spent the majority of the trip completely covered up or hiding in the shade, scowling at everyone and clearly having a terrible time.

My favorite anecdote is the moment in Iceland (definitely the country) when, after a three-hour bumpy bus journey across lunar-looking landscapes, we arrived for a swim at a secret lagoon. I stood in wonder before a milky blue steaming pool surrounded by snowy peaks that looked like it was on another planet, just as the journalist next to me declared that she couldn't possibly have a dip because she'd blow-dried her hair that morning. It reminded me of the blogger who announced—on the first night of a pasta-tasting tour around Italy—that she didn't like pasta.

When I first started doing press trips, there would occasionally be a token "blogger" who other journalists would slightly look down their noses at as if they were a sub-species. Now influencers get their own trips and can be paid thousands of pounds for posting a single image from it. Occasionally, publicists mix influencers and journalists together, and the results can be a real eye-opener into what it takes to gain and maintain huge numbers of followers.

I've been on a spa trip with fitness bloggers who documented every second—including filming on their phones inside the changing room and during their treatments. "Hi guys, soooo I'm here having my massage right now…" I've watched a French fashion blogger drag her duvet onto the beach at sunset (aka "golden hour"), then ask a waiter to take endless pictures of her writhing around in her underwear. I've seen an influencer order not one but three "floating breakfasts" (for those who like their eggs to come on a tray that hovers on the surface of their swimming pool) to be photographed by drone and then not eat a single mouthful. I've witnessed a well-known Italian fashion blogger take "candid" selfies that involved full hair and makeup, a stylist, a wind machine, a professional photographer, and several lights and reflector boards.

These influencer moments felt surreal, but perhaps it's just the extreme end of the lengths we all go to in order to present the best side of our travels—and our lives—on social media. Perhaps we've always curated an idealized version of our holidays—postcards rarely depict that drizzly day or the beach that's cluttered with litter. And although group press trips aren't always fun, and every time I go on one I swear it's the last, you are still in an amazing place having a kind-of-amazing time. And, of course, no one wants to hear you complain about your "free holiday," so it's only with other travel journalists that you can really let rip about your press trips from hell.

All travel journalists have many, many tales like this and, while press trips have changed over the years in shape and budget, I'm pretty sure they've been happening in some form for centuries. Who knows, maybe an innkeeper offered Egeria a free night's stay in return for a favorable report in her letters? But one of the earliest and most interesting press trips has to be the record-breaking journey undertaken by American journalist Nellie Bly in 1889. Although there were no publicists and no influencers back then, her itinerary made headlines around the globe.

While working at Joseph Pulitzer's newspaper, the *New York World*, Nellie came up with a "proposal to girdle the earth" in an attempt to outpace Jules Verne's fictional eighty-day itinerary. Like most of the press trips I've been on, with just two days' notice, she threw some underwear and a tub of moisturizer in a bag, hopped aboard a steamship for London, and didn't stop moving for seventy-two days. Unlike any press trip I've been on, she returned home to thousands of fans and a gun salute. The stunt made her the most famous reporter in America.

As with many of the women in this book, Nellie's life story would make a great movie. Born Elizabeth Jane Cochran in 1864 to a poor family in western Pennsylvania, she was one of fifteen children and had little formal education. And yet she became one of the most celebrated journalists of her time. Her work changed public policy, her outfits influenced fashion trends, and her adventures inspired board games. She even made a cameo in *The Great Gatsby*—the character of Ella Kaye, a tough newspaperwoman, is said to be based on Bly.

Nellie's career began when she was still a teenager, when she wrote an anonymous letter to her local paper, the *Pittsburgh Dispatch*, in response to a patronizing article titled "What Girls Are Good For"—according to the author it was basically having babies and keeping house. Elizabeth's blistering letter pointed out all the social advantages doled out to boys not girls, and yet: "Girls are just as smart, a great deal quicker to learn; why, then, can they not do the same?" The editor asked the writer of the letter to identify themselves and went on to hire her as a columnist, and she began writing under the pen name Nellie Bly.

Although she spent many years in the so-called "pink ghetto" of journalism, assigned stories about "women's subjects" such as gardening and fashion, Nellie eventually got tired of writing about flower arranging and moved to Mexico to be a foreign correspondent. When she arrived back in New York in 1887, it was the beginning of the "yellow journalism" movement, when sensationalized headlines massively increased circulation. Between 1870 and 1900, the number of newspapers sold each day rose almost sixfold, and Nellie was there to ride the wave.

She quickly made her name by going undercover in a mental asylum pretending to be "positively demented." Her article "Ten Days in a Mad House" exposed the shocking brutality in such institutions and led to reform in mental healthcare. She became one of the first "girl stunt reporters," a tradition that continues to this day and basically involves sending a female reporter—and it usually still is a woman—to try something strange or risky or weird. In my time, I've been injected with vitamins, made a hat out of a dead pigeon, and spent a week running a bed and breakfast in rural Northumberland. But that is small fry compared to Nellie.

Along with her ten-day stay in Blackwell's Island Insane Asylum, she trained with a boxing champion, visited an opium den, posed as an unemployed maid, and dabbled in elephant training and ballet. Once, to expose the workings of New York's white slave trade, she even bought a baby. But her most well-known piece of journalism was her race around the world.

Nellie, then twenty-five, came up with the idea for her globe-trotting scoop on a Sunday afternoon where she wished she could be "at the other end of the earth." Although her editor liked it, the paper's business manager didn't agree. As Nellie recounts in her article: "'It is impossible for you to do it,' was the terrible verdict. 'In the first place you are a woman and would need a protector, and even if it were possible for you to travel alone you would need to carry so much baggage that it would detain you in making rapid changes. Besides you speak nothing but English, so there is no use talking about it; no one but a man can do this.' 'Very well,' I said angrily, 'Start the man, and I'll start the same day for some other newspaper and beat him.'"

So Nellie began her 40,070-kilometer journey, traveling with only the dress she was wearing ("a plain blue broadcloth and a quiet plaid camel's-hair...the most durable and suitable combination for a traveling gown"), a silk bodice, a sturdy overcoat, several changes of underwear and a small travel bag carrying her toiletries. "Packing that bag was the most difficult undertaking of my life; there was so much to go into such little space," she wrote. "One never knows the capacity of an ordinary hand-satchel until dire necessity compels the exercise of all one's ingenuity to reduce everything to the smallest possible compass. In mine I was able to pack two traveling caps, three veils, a pair of slippers, a complete outfit of toilet articles, ink-stand, pens, pencils, and copy-paper, pins, needles and thread, a dressing gown, a tennis blazer, a small flask and a drinking cup, several complete changes of underwear, a liberal supply of handkerchiefs and fresh ruchings and most bulky and uncompromising of all, a jar of cold cream to keep my face from chapping in the varied climates I should encounter." I now think of Nellie every time I'm faced with the task of decanting my toiletries into 100-milliliter bottles at airport security.

Nellie's formidable, no-nonsense approach to packing runs throughout her writing. "I always have a comfortable feeling that nothing is impossible if one applies a certain amount of energy in the right direction," she wrote. "When I want things done, which is always at the last moment, and I am met with such an answer: 'It's too late. I hardly think it can be done'; I simply say: 'Nonsense! If you want to do it, you can do it. The question is, do you want to do it?'"

Just like me, Nellie loved eating ("There is nothing like plenty of food to preserve health.") and also a lie-in. As someone who has been known to snooze for twelve hours straight, regularly naps, and often awakes in hotels to find that breakfast has long been cleared away, I admired her unashamed laziness. She complained that her boat out of New York—the *August Victoria* steamer—left at the unholy hour of 9:40 a.m. "If some of these good people who spend so much time in trying to invent flying machines would only devote a little of the same energy toward promoting a system by which boats and trains would always make their start at noon or afterward, they would be of greater assistance to suffering humanity."

She was constantly napping at inopportune moments and when people tried to rouse her she remarked, "I generally get up when I feel so inclined"—something I would quite consider tattooing on my eyelids. While on a steamer, she woke up to find the captain in her cabin. "'We were afraid that you were dead,' the captain said when he saw that I was awake. 'I always sleep late in the morning,' I said apologetically. 'In the morning!' the Captain exclaimed, with a laugh, which was echoed by the others, 'It is half-past four in the evening!'"

On her train ride from London, she refused the requests of her guide to enjoy the scenery. "Honestly, now, I care very little for scenery when I am so sleepy." Like me, she also got travel sick, but Nellie was much better at describing her "lively tussle with the disease of the wave." She wrote about how, at one point, she leaned over the side of the ship, "giving vent to my feelings. The smiles did not bother me, but one man said sneeringly: 'And she's going around the world!'"

Like many journalists forced to endure a press trip, Nellie could often be hilariously jaded, complaining that, in Aden, Yemen, she found "nothing extraordinary." "Nearby was a goat market, but business seemed dull in both places," she wrote. In fact, Aden was just "dirty, uninviting shops and the dirty, uninviting people in and about them." Nellie's opinion of Kandy in Sri Lanka is that it was "pretty but far from what it is claimed to be." When traveling on a cold train, she regretted not being held up by bandits, as at least that would have been some excitement to make her "blood circulate." While writing about Hong Kong, she uses the word "dirty" five times in a single sentence.

But she made some beautiful observations, too. Sailing boats emerging out of the darkness in the Suez Canal are "like moths flocking to a light;" rubies in Ceylon are "like pure drops of blood." Although for the modern reader her racist comments about "the natives" are jarring, she can also be empathetic and kind. Disembarking the ship at Port

Said in Egypt, when her fellow passengers armed themselves with canes and parasols to "keep off the beggars," she refused, saying, "a stick beats more ugliness into a person than it ever beats out." (Although she ruins it slightly when boat-taxis fight with each other to carry her, exclaiming, "probably there was some justification in arming one's self with a club.")

She noted the strong connection you get with fellow travelers when you're solo. "Had I been traveling with a companion I should not have felt this [bond] so keenly, for naturally then I would have had less time to cultivate the acquaintance of my fellow passengers," she wrote. She also experienced the difficulties every solo traveler feels occasionally—the pain of leaving new friends and a place you've only just come to know and being "once again alone in strange lands with strange people."

But Nellie wasn't just racing round the world to beat Phileas Fogg's time. Unbeknownst to her, *Cosmopolitan*—then a New York newspaper, not a magazine known for sex tips—sent their own reporter, Elizabeth Bisland, to travel around the world in the opposite direction of Nellie to try to beat her. Nellie was more than halfway through her journey and about to board a ship for Japan when she realized that she had a real-life competitor. "I am running a race with time," she insisted to the ship's officer, who had told her she was going to lose. "Time? I don't think that's her name," he replied.

The race element was just one of the ways in which public interest in Nellie's quest was sustained. The *World* also organized a "Nellie Bly Guessing Match" in which readers were asked to estimate Nellie's arrival time to the second with the grand prize consisting of a free trip to Europe. More than 900,000 people entered. The paper also published a full-page Nellie Bly board game, based on snakes and ladders, with instructions such as "Suez Canal, Lose a Throw" and "Indian Mail Accident, go back 5 days."

During the final leg of her trip—a train from San Francisco to New Jersey—Nellie was met with crowds of well-wishers at every stop, offering flowers and sweets and trying to touch her hand. "It was glorious! A ride worthy of a queen," she wrote. She arrived back in New York with a sunburnt nose and newfound fame, having traveled alone and "fancy free" for almost the entire journey.

This "intrepid petticoated traveler" had also set a new world record—for anyone, male or female—to go around the world and no doubt made travel, especially solo travel for women, feel more accessible. Her distinctive traveling outfit—a cap, high-necked blue jacket and skirt, and a long wool tweed coat—became so popular that women copied the look for more than a decade.

But Nellie was only the first of many female journalists who began scouring the globe in search of a story. Perhaps the most well-known is Martha Gellhorn. I'm sheepish to admit that until I started researching this book I only knew of her for exactly the reason she didn't want to be remembered—as Ernest Hemingway's third wife. And yet when I read Martha's 1978 memoir, *Travels with Myself and Another* (the "Another" is Hemingway), I was fervently underlining every other sentence.

She noted, "The only aspect of our travels that is guaranteed to hold an audience is disaster," and goes on to recount her best "horror journeys," which make my press trip tales sound like, well, a holiday. She strode, mostly alone, through fifty-three countries, made homes in nineteen different places, and described herself as feeling "permanently dislocated—*un voyageur sur la terre.*" Like many travelers, she was "enamoured of surprises and excitement and jokes and risks and odd people."

In a journalism career that spanned sixty years, she is probably most celebrated for her fearless war reporting—she covered almost every major conflict of the twentieth century. She stowed away on a hospital ship in a nurse's uniform to report on the D-Day landings—the only woman to cover the Allied invasion. She was also one of the first journalists to witness the liberation of Dachau concentration camp in 1945.

Her marriage to Hemingway lasted five years. While she was reporting on the Blitz in London he sent her a cable saying, "Are you a war correspondent or a wife in my bed?" and they divorced soon after. In all future interviews, she banned his name from being mentioned. "Why should I be a footnote to someone else's life?" was her reasoning.

Martha worked as a journalist until well into her eighties. In the spring of 1990, when she was eighty-one, she went door-to-door in Chorrillo, a slum of Panama City, reporting on civilian casualties resulting from the U.S. invasion a few months earlier. She was living in London, aged eighty-nine, when she was given a terminal cancer diagnosis and took her own life with a cyanide pill shortly afterward.

Like Martha, Nellie also had a notable marriage. When she got back from her round-the-world jaunt, at the age of thirty-one, she married a seventy-three-year-old millionaire. She retired from journalism to run his steel company and became the inventor of a new type of milk can and a stacking rubbish bin. After her husband died, she went bankrupt and so she returned to journalism. Just like Martha, Nellie became a war correspondent, reporting from the Eastern Front during the First World War.

In her fifties, Nellie became the first woman, and one of the first foreigners, to enter the war zone between Austria and Serbia. She was arrested by German soldiers who mistook her for a British spy. In one dispatch for the *New York Evening Journal*, under the headline "Nellie Bly in the Firing Line," she described the last hours of a wounded Russian soldier: "'Could Emperors and Czars and Kings look on this torturing slaughter and ever sleep again?' I asked the doctor. 'They do not look,' he said gently." When the war ended, Nellie returned to New York, where she died of pneumonia aged fifty-seven. Before Nellie Bly, female journalists, let alone female war correspondents, were few and far between.

In the final sentence of her report for the *World*, Nellie Bly recognized the real joy of her adventures, or the press trip, or Gellhorn's "horror journey," and maybe all travel—that it makes you glad to be home and able to appreciate your home with fresh eyes. Solo travel doesn't just expose you to new experiences, it changes how you see the world by making the familiar strange again.

For the first few days back in London after a trip, it's as if someone has peeled away a layer of my eyeballs. The sky looks brighter, the buildings shinier. Graffiti I walk past every day isn't graffiti, it's street art. As Nellie put it: "I took off my cap and wanted to yell with the crowd, not because I had gone around the world in seventy-two days, but because I was home again."

HOW TO RECORD YOUR TRAVELS

When you're traveling alone, experiences can often feel more profound and intense. When there's no one there to share them with, it's even more important to record them in some way. It's also surprisingly enjoyable to relive a trip after time has passed, even the parts you think you'd rather forget.

1. Sound it out.

It's funny that we're so obsessed with taking pictures on holiday but rarely think about recording sounds. It's something I've only started doing relatively recently. I think for a long time I was put off the idea because I went on a press trip to Morocco with a man from Jazz FM who insisted on stopping the bus every few miles so he could get out and record the traffic for "color." But I've now recorded some great aural memories to listen to later—girls skipping in Mexico City, a busking flamenco performer in Seville. It can also feel less intrusive to record sounds on your phone than take a photo.

2. Zero in on the details.

Travel can feel like a constant battle between wanting to take photos of every single thing and also be "in the moment," experiencing what's in front of you rather than through the medium of an iPhone screen. One friend decides on just one thing she will photograph each trip and keeps to that. When she went to Japan on her honeymoon, she took a photo of every bed her and her husband slept in. Just the beds. I know someone else who documented a road trip around California by snapping every burger from the In-N-Out takeout variety to the ridiculously expensive wagyu beef patty she ate in Las Vegas. I like to remember to take photos of the things that aren't so beautiful or Instagrammable. The grotty retro bathroom. The comically misspelled menu. These things bring instant joy when you look at them later in a way that 800 photos of the Grand Canyon probably don't.

3.   Try journaling.

No matter how tired I am, I try to scribble down a few words at the end of every day. Sometimes they are nonsensical, and I manage nothing more profound than "hot, temple, monkey, bus." Other times I think I don't have much to say but end up writing pages, just enjoying the pen flowing and the words appearing. I don't actually tend to use these notes again when I'm writing up a trip for my work, but I like to think the act of writing somehow cements what's happened that day in my brain.

4.   Record what you see.

My friend Emma is a wonderful illustrator who, when she's traveling, draws a daily journal of her escapades. When I traveled in India with Emma, we had many sixteen-hour bus journeys. I spent them vomiting into a plastic bag or sleeping. She spent them drawing incredible pictures of the people on the bus. I'm not sure how our friendship survived this inequality. But if you're not a fan of drawing or writing, there are other ways to journal. I know a few people who, while they're traveling, save receipts and tickets and menus, and when they're home they make collages from them. Or why not try making a video with the One Second Everyday app, which encourages you to record just one second of footage every single day and then threads them together into a movie of your trip.

5.   Tell the folks back home.

Whether it's via social media or just on round-robin emails, or even—full retro points—a postcard, this can be great training if you want to get into travel writing or have plans to pen your memoirs at some stage. Some of the best advice I know about finding your voice as a writer is just to write like you would talk to your friends. One editor I had early in my career told me that the key to cracking an intro to a feature was to imagine what you'd start with if you were telling your friends the story down at the pub. But maybe with fewer swear words.

CHAPTER FOUR
The Women's Movement

Around the world on the "freedom machine" with Annie Londonderry

At 11:00 a.m. on Monday, 25 June, 1894, hundreds of suffragists, friends, family members, and rubbernecking passersby on their way to a baseball game gathered at the Massachusetts State House in Boston. They were there to witness the start of something that had never been attempted before. This was another around-the-world trip by a woman—but this time she was on a bicycle!

Annie Cohen Kopchovsky seems like a pretty unlikely candidate to attempt this historic journey. A Latvian immigrant to Boston, she was twenty-three and had three young children at home. Oh, and she'd never ridden a bicycle before. No doubt inspired by Nellie Bly's attention-grabbing feat just five years before, Annie also wrote articles about her trip—one of her by-line names was even Nellie Bly Junior. In a picture of her taken before she set off, she looks like Miss Gulch from *The Wizard of Oz*. Hair scraped back in a severe bun, tailored jacket with Victorian leg of mutton sleeves, a straw bonnet, and a steely gaze contemplating the miles ahead of her.

Carrying only a change of underwear and a pearl-handled revolver, Annie left Boston on a weighty forty-two-pound Columbia bicycle dressed in long skirts and didn't come home again for over a year. It was a trip that, as she described it, was filled with harrowing adventure, frequent danger, and endless drama. Over the course of it, she went from anonymous working mother to global celebrity.

Annie claimed that the trip was inspired by a wager between two wealthy Boston sugar merchants who bet that no woman could "traverse the globe on a wheel." In reality, she probably just concocted the idea as an enterprising way to escape her life for a bit. But the conditions of the bet, as she described it, were that she was allowed just five cents a day for expenses and was obliged to earn $5,000 along the way. If she succeeded, she claimed she would win $10,000 in prize money. Considering the average annual salary for an American in 1895 was around $1,000, these were huge sums.

In perhaps the first ever example of a female athlete getting sporting sponsorship, Annie funded her trip by letting companies advertise on various parts of her body, turning herself into a moving billboard. She even changed her name from Kopchovsky to Londonderry, after a mineral water brand offered her $100 to carry their placard on her bicycle and adopt their name. Although as Annie's great-grand-nephew, Peter Zheutlin, asserts in his book *Around the World on Two Wheels*, the name-change was about more than sponsorship money. It was not only a more memorable alias, but it also disguised Annie's faith at a time when not all parts of the world welcomed Jewish people.

When I first came across Annie while researching solo female travelers, she seemed like a kindred spirit. Obviously I felt an affinity with someone who would undertake such an ambitious trip with little or no expertise or advance planning. I've also had some of my best adventures by bicycle. I didn't pass my driving test until I was thirty-four, so for most of my traveling life two wheels had been my main mode of transport. But outside of the necessity of cycling, I've always loved the pace and the simplicity of it. You need very little fitness or coordination to master it. It feels empowering to be propelled by nothing more than your own energy, feeling the air and the scenery and the world around you, moving with you, without any separation from it.

I've come to believe that pootling along in the saddle is the best way to see a city—gliding serenely past sights, stopping when something catches your eye. Why would you want to be trapped in a car when you could feel the breeze in your hair? I've ridden electric bikes through sugar plantations in Mauritius, rented a rickety old bike in India that promised to give up on me every time we went up an even remotely steep hill, and whizzed through Puglian olive groves on a mountain bike. But it hasn't always been plain cycling. It was even a bike accident that precipitated me and Sam getting together.

We met while studying abroad together in California. I'd flown to LA in 2005 with a beaten-up copy of *On the Road* and big dreams. I remember landing in LAX and being mesmerized by the size of everything, from the coffees to the cars to the people. Everything seemed bigger, brighter, brasher. I felt like Dorothy stepping into the technicolor world of Oz. Toto, we're not in Surbiton any more. I got in a taxi to my hostel and stared at the uncannily familiar green road signs advertising Sunset Boulevard and Ocean Drive and felt like I was living out all my pop culture fantasies. When I got to Santa Barbara—a moneyed beach enclave where the houses and streets and people were so perfect it felt like living in a film set—it was everything I had imagined the Californian dream to be.

UC Santa Barbara was known affectionately by students as the University of Casual Sex and Beer and for me it was every American high school movie brought to life. I played beer pong and drank out of red cups (*American Pie*). I pledged a sorority (*Legally Blonde*). I was asked to read aloud in every English class (*Grease 2*). I had planned to fall in love with a hunky surfer dude and have a few blond babies and live in a Malibu beach house forever. But after a few dalliances with boys named

Jason and Brad who had chests so hard it was painful to rest my cheek on them, a geeky Jewish boy from west London caught my eye.

We met in one of my English classes where I was annoyed every time he spoke because I wanted to be the only person who everyone thought had a "cute accent." Not only that but he seemed to be the smartest by miles and could not only quote Nietzsche but pronounce his name correctly, too. We started hanging out after class and going to parties together and laughing about Americanisms. Sam suggested we write a column for the student newspaper about being Brits in California. You know that meme about creative people having 7,564 browser tabs open? That was Sam, flicking between Shakespeare and Snoop Dogg via *The Sandman*. It was hard to keep up.

Santa Barbara was full of bike paths, and we used to cycle around together—Sam impressed me by riding his bike with no hands *and* rolling a cigarette at the same time. Perhaps that's what inspired me to think that I'd be absolutely fine to ride my bike from Sam's dorm room back to my own little apartment in the dark after a microdot of acid. The dot *was* very micro, after all. Little did I know that drugs and cycling do not mix (this was years before Lance Armstrong).

All I remember about the crash is coming off my bike at full speed and landing face-first on the ground with my hand still clutching my iPod. I've subsequently heard that women have an innate instinct whereby if they happen to fall over while they're holding something, they will keep on holding it no matter what, in case said thing is a baby. I don't know if that's true, but I do know that I apparently cared more about my Apple product than I did my face. The iPod had not a scratch on it, yet I seemed to be face down and eating a curb sandwich.

Unsure of what to do but aware that I was lying in a pool of my own blood and that my mouth felt very, very weird and knowing it might be a while before someone cycled past, I shakily got up, got back on my bike—there's a reason that's such a good saying—and rode home to my roommates. Through the haze of bud they took one look at my scraped-raw, already-swollen face and my teeth (which were now sticking out at funny right angles like piano keys) and took me straight to the emergency room. There, a kindly doctor tried (and failed) to just "push them back in there, kiddo!" before giving me stitches in my lip, a Vicodin prescription, and a surgery appointment the next day.

I'd never broken a bone before and apparently your jaw is the hardest bone in your body to break. So I felt both strangely proud that I'd gone straight to the top and also pretty lucky that I hadn't landed on my skull. I spent the next six weeks with my jaw wired together (like Lisa in that *Simpsons* episode) with black Frankenstein-esque stitches across my lips, high as a kite on prescription painkillers, unable to eat solid food, and existing solely on banana and peanut butter milkshakes.

It was under these deeply sexy circumstances that Sam and I had our first kiss (of sorts—it turns out it's not that easy to snog with a wired jaw). Although no doubt there had already been moments of romantic tension long before he came over every day to nurse my injuries. The truth is, I had fancied him for ages, but it was a bike crash that brought us together. And I still have a scar on my upper lip to remind me.

Annie Londonderry got into her fair share of scrapes, too. This was the 1890s, so cycle lanes weren't a thing, nor were pavements for the most part, so she often had to push her unwieldy, heavy bicycle along sand and gravel roads in the heat of summer—and in full skirts. The trip was so exhausting that when she arrived in Chicago after a thousand miles and three weeks on the road, she thought about giving up the whole thing. "When I left Boston...I rode a forty-two-pound wheel and was attired in skirts. The result was that when I reached Chicago, I was completely discouraged," she told reporters. A new bicycle—a light, twenty-pound Sterling men's frame that was ivory with a gold trim—and a change into bloomers convinced her to carry on.

It was actually the popularity of cycling for women that led to the shift in fashion known as the dress reform movement. Restrictive skirts and long-sleeved shirts with collars were out in favor of something more practical and cycling-friendly. In 1891, the British suffragette and writer Helena Swanwick point-blank refused to wear a skirt again after a bike crash. "It is an unpleasant experience to be hurled onto [the ground] and find that one's skirt has been so tightly wound around the pedal that one cannot even get up to unwind it," she wrote. The solution? Cycling bloomers, or a "bicycling costume," as it was known at the time—trousers that fit tightly below the knee. Cool fact—this is where we get the term "pedal pushers" from (I personally had a particularly fetching denim pair in the 1990s that made me look like I was in the pop group B*Witched).

Up until the 1880s, "bone shakers" or "ordinaries" as bicycles were known were considered solely the domain of men. But the invention of the "safety bicycle," which meant the rider could put their feet on the ground when stationary, suddenly made cycling more accessible for women and children. Although in the early days it took considerable courage for women to defy disapprobation and endure being stared at, jeered at, and even pelted with vegetables.

Then there were the medical concerns about women getting in the saddle. That bike riding might be sexually stimulating for women was a real concern in the 1890s. So-called "hygienic saddles" began to appear, featuring seats with an open space where a woman's vulva would normally make contact. Several unsuccessful attempts were made to produce a bicycle with both pedals on the same side in order to allow women to ride side-saddle and avoid any unwanted excitement. All kinds of arguments were dreamt up to prevent women from cycling, with claims that the vibrations could rattle your innards and lead to everything from infertility to promiscuity. Most worrying of all, was "bicycle face" and the idea that the tense expression of concentration required to ride a bike would give you wrinkles. *Harper's* magazine recommended chewing gum to keep the effects at bay.

Dame Ethel Smyth, a composer, recalled that when she took to two wheels around 1890, most of her female relatives were horrified by her indelicacy. Helena Swanwick wrote that, around the same time, when she and her husband would ride around Manchester, she was harassed by mill hands, bus drivers, and cabmen whose normal deference to women of her class disappeared in the face of her "boldness." From my own experience, not much has changed. Riding through London on a bike, I've experienced more catcalling and harassment than I get on foot.

For many people, women on bikes were not just an oddity or an indecency, but a threat to civilization. The cycling magazine *The Bearings* had this to say about female bike races on July 25, 1895: "The spectacle of…females straining every muscle, perspiring at every pore, and bent over their handle-bars in a weak imitation of their brothers is enough to disgust the most enthusiastic of wheelmen." Women kept pedaling, regardless. Between 1891 and 1896, it is estimated that the number of female cyclists grew between 100 and 400 times, with 3.2 million female cyclists in the United States, Great Britain, France, and Germany.

No matter what many disapproving men (and many women) thought about it, cycling was now a mass phenomenon, and bike manufacturers used images of female freedom in their adverts. Columbia advertised its bikes with a nude woman balancing a glass of wine on her head sitting on one of its bikes, which is something I'd like to try sometime. Another depicted a woman climbing a hill astride her Columbia while her male companion wipes his brow with a handkerchief and struggles to push his bike up beside her. Women in cycling posters were confident, in control, and out in the world on their own terms, and Annie Londonderry became the living embodiment of this image.

When she was unable to find a hotel or rooming house, Annie slept under bridges and in barns. At points, she only ate apples and "became accustomed to relieving oneself in creative ways and not always in pleasant places." Was she scared and did she have the 'Oh fuck, what have I done?' feeling? Or was she just excited to be liberated from life in a Boston tenement with three toddlers under five? Probably a bit of both.

As her journey went on and as public fascination with her grew, Annie played down the fact she had a husband and three children back at home. People presumed that she must have been single to attempt such a trip and, at one point, when asked what she would do with her winnings, she even quipped, "Why, I'll marry some good man and settle down in life." According to one newspaper report, she received "147 proposals of marriage from men of wealth and nobility."

Who knows what the public in the 1890s would have made of a woman leaving behind not only her husband but a five-year-old, a three-year-old, and a two-year-old. Even today, that would no doubt cause raised eyebrows and a few screaming headlines, given that only 7 percent of British adults believe women with pre-school children should have full-time jobs. Our expectations of a woman's place haven't moved that far outside the home.

Unsurprisingly Annie (and her bike) became a poster girl for "the New Woman"—a *fin de siècle* archetype who was demanding really outlandish things, like an education, enfranchisement, and the right to pursue a career. For women at the turn of the century, the bicycle represented more than just a way to get around—it was a vehicle for freedom. Before the bike, middle- and upper-class women were meant to stay indoors, and if they did venture out, it was with chaperones, usually to what were deemed "acceptable" public spaces. If they were permitted to be on the move, it was slow and dainty and controlled (walking or in carriages), not freewheeling on bikes.

It's hard to imagine now just how much the bicycle broadened horizons for women or gave them a sense of movement and mobility after a lifetime of cumbersome skirts and constricting corsets. Susan B. Anthony, the American women's rights activist, called the bicycle "the freedom machine" and told Nellie Bly in an interview in 1896 that cycling "has done more to emancipate women than anything else in the world." The bicycle became the symbol for women's liberation.

There's an extraordinary black and white photograph from 1897 that shows Cambridge students protesting against the admission of women to their university by hanging up an effigy of a woman on a bike. In the 1910s, suffragettes would ride around on bicycles with "Votes for Women" banners streaming off the back, and they even rode them to commit arson attacks. In 1912, they blocked then–Home Secretary Winston Churchill's motorcade with their bicycles. The women's suffrage movement even had its own bike. In 1909, an advertisement for a purple, green, and white model, with a drop frame to accommodate long skirts appeared in the pages of the magazine *Votes for Women*.

Like every female athlete both then and now (as well as female musicians, actors, and public personas of any kind) Annie's appearance was frequently commented on. Both favorably (The *Elkhart Daily Truth* commented that Annie's bloomers outfit "exquisitely becomes her petite figure") and unfavorably (*Le Figaro* reported that she was "Mannish... with a bony face"). Some French newspapers even cast doubt on whether she was a woman at all. But Annie turned this scrutiny around and used it to encourage more women to get on their bikes. "If women will exercise properly on a wheel they will have nicely rounded figures, bright eyes, and healthy cheeks," she told one reporter. "My work on the bicycle since I started on this trip has developed me wonderfully."

From the start, Annie was a pro at manipulating the media on what she called her "whirl around the world." She coined catchy names for herself such as the "circum-cycler" and the "lone rider." The *Buffalo Courier* patronizingly described her as a "clever and intrepid little wheel-woman" and her journey "one of the most perilous and remarkable trips ever undertaken by a woman." She had some sassy comebacks for those who questioned her. A report in the *Rochester Post-Express* details a police officer encountering Annie and demanding an explanation for her outlandish appearance ("an ugly looking riding costume and...muddy shoes"). She replied, "My autographs cost twenty-five cents apiece. If you want one I should be glad to give it to you."

While in bicycle-mad Paris, Annie became the object of intense attention and speculation. "I found out what they liked and gave them plenty of it," she said of Paris, but that could easily apply to her attitude on her trip as a whole. Throughout her adventure, Annie was creative with the truth. No evidence has been found of any wager made back in Boston about her journey, and the terms of this alleged bet changed all the time.

At best, Annie was a fantasist who loved spinning tall tales; at worst, she was guilty of willful deception and fraud. At various points, she claimed to be an orphan, a wealthy heiress, a medical student, and the cousin of a U.S. congressman (none of which were true). Although she often took trains and boats as part of the trip, she only revealed this when specifically asked.

According to Annie, she was held up by highwaymen in France, broke her wrist in Iowa, and contracted pneumonia in Denver. She apparently rode through Alexandria, Port Said, Jerusalem, and Siberia. In India she allegedly hunted Bengal tigers. She was caught up in the Sino-Japanese war in 1895, fell through a frozen river, took a bullet in the shoulder, and was thrown in a Japanese prison.

Even if she had cycled nonstop, it would've been impossible for her to cross India and cycle overland to the China coast in seven weeks. Annie was clearly a subscriber to the Mark Twain school of travel yarns: never let the truth get in the way of a good story. Which I suppose is the other benefit of going it alone—there's no one to say it didn't happen. By the time Annie arrived in Yokohama in early March, skepticism about her claims was, justifiably, on the rise. According to the *Japan Weekly Mail*, she was merely "traveling the world *with* a bicycle." She was accused of shameless self-promotion. "Glib and vulgar is Annie Londonderry," wrote *Cycling Life*. "Her aim is to be notorious."

But clearly, from Annie's point of view, the more column inches she got, the better. Like an early version of today's social media influencers, at one point she stage-managed a photo shoot with bandits accosting her at gunpoint. She sold photographs of herself, which she autographed, to pay for her travel expenses. She was savvy and way ahead of her time when it came to understanding the power of the press. Fame, money, and an escape from the ordinary were Annie's real goals. The whole cycling around the world thing merely came along for the ride.

Like Annie, I didn't get on a bike for the health benefits of physical activity. But it was another trip to California in my late twenties (an actual trip, not a hallucinogenic one) that taught me the life-changing magic of moving around. Despite being the sort of person who seemed allergic to any kind of exercise, I accepted a commission to report on fitness trends in LA. I spent a week learning circus skills, joining a street-art spotting jogging squad, and dodging laser guns while running over boulders in something that resembled Laser Quest but without the dry ice. I cycled on a stationary bike in a candlelit room while a woman who was 90 percent abs screamed at me about my inner essence.

I also took part in something called "The Class" where around forty Lululemon-clad women went into a room for what I can only describe as scream therapy mixed with HIIT. A lady in a crop top with a Britney mic stood at the front of the class, shouting motivational phrases like, "You are not your Instagram. You are not how your hair looks. You are not what they say you are," while we did burpees and screamed very loudly. One exercise involved thumping your body à la Matthew McConaughey in *The Wolf of Wall Street* and grunting. Another had me doing squat jumps and roaring. I loved it. And it gave me more of an insight into LA than any tour of the Hollywood Hills or visit to the Getty Museum ever could.

Something changed in me in that mad, exercise-fueled week. I went from thinking sport was something to be avoided at all costs to realizing it was like a free, socially acceptable form of drug consumption. Food tastes better. You have more energy. You need less sleep. Your anxiety gets burned away. I was hooked.

Although I didn't then have a Damascene conversion to running ultramarathons, now when I'm traveling I often undertake the sort of exercise I'm far too lazy to even contemplate at home. I've done sunrise yoga on a jetty overlooking the jungle in Bali. I've been kitesurfing in Essaouira, Africa's "windiest city." I've been coasteering off the Isle of Man (to the uninitiated, this involves donning a wetsuit and a helmet and flinging yourself off some rocks and then scrambling back up others). And let me just reassure you that I am absolutely not a "sporty" person.

At school I did everything I could to bunk off PE (I was constantly on my period for about seven years), and when I did turn up I was picked last for every team. To make matters worse, because I was tall, PE teachers were convinced that I must be good at netball or high jump or something. Yet I had the coordination of a politician on *Strictly*. Because I was so bad at it, and because I was so unfit it made me feel sick to run around the track as "a warm up," I loathed physical exertion and saw no reason why I should waste any effort getting a stitch.

This continued well into my twenties. The only trainers I owned were Converse, and if I was ever forced to do something vaguely physical, like running for a bus on the way to the pub, I felt like I was having some sort of stroke. So it was fairly revelatory to me that exercise could be (a) enjoyable and (b) that I'd actually pay my own money to do it.

I now actively seek out ways to move my body on a trip. Not only is it the best jet lag cure I've found (it somehow manages to both tire you out if you need to sleep and wake you up if you're groggy) but doing something active becomes a kind of souvenir in itself. There's a theory that memories are kept in our tissues and "the body keeps the score," and often I'll be in pigeon pose in a yoga class in London and hear the voice of the teacher I had in Sweden, who called it dove pose. Or I'll be cycling along and something about the turn of the pedals under my feet and the position of my hands on the handlebars will take me right back to sunny days in Santa Barbara, even if I'm just breathing smog on Old Street roundabout. I agree with the Beat poet Gary Snyder who said, "That's the way to see the world: in our bodies."

Although Annie's odyssey made headlines around the world, she was certainly not alone in traveling impressive distances on two wheels. In 1893, sixteen-year-old Tessie Reynolds rode from Brighton to London and back wearing knickerbockers. Around the same time, husband and wife team William and Fanny Workman wrote several books about their travels on bicycles to far-flung destinations such as Algeria and India. The glamorously named Elizabeth le Blond (1861–1934), a mountaineer and expert in snow photography, became a competitive alpine cyclist.

There is doubtless an element of Annie's freewheeling spirit—and her minimal packing list—in Dervla Murphy, who cycled from Ireland to India in 1963 on a bicycle with no gears armed with little more than a pistol and a journal. More than fifty years and twenty-six travel books later, she has traveled to Afghanistan, Peru, and Siberia (among others), usually on her bike named Roz or sometimes by donkey or mule. She's broken ribs and ankles, contracted hepatitis and malaria. In Ethiopia, she was robbed three times, once by armed bandits who nearly killed her. In the Balkans, she was set on by wolves. In Tibet, she wore the same clothes, day and night, for three months (and ended up with body lice). "Well, I'm quite simply enjoying myself," she told an incredulous Sue Lawley who interviewed her for *Desert Island Discs* in 1993 and wanted to know why on Earth she did it.

Less well known but no less inspiring is Anne Mustoe, the Suffolk headmistress who cycled around the world in 1987, wrote several books about her adventures, and died in 2009 while riding her trusty Condor bike through Aleppo. There's also the adventure cyclist Josie Dew who, in 2007, cycled 3,000 miles with her baby daughter strapped to the back of her bike. Jenny Graham, the Scottish ultra-endurance cyclist, would've had no need for Annie's tall tales. In 2018, she became the fastest woman to have actually cycled all the way around the world unsupported when she set off from Berlin and circumnavigated the globe in 124 days. More recently, I've been inspired by the adventures of activist and cyclist Kate Rawles, who rode 8,288 miles across South America on "Woody," a self-built bamboo bicycle.

Annie's trip ended quietly. Fifteen months and thousands of dubiously cycled miles later, she was home, with her left arm in a sling and "cheeks bronzed from exposure." There were no parades or fanfares like Nellie Bly received, but news that her adventure had finished was reported as far away as Milan and Honolulu. There's no record of Annie's reunion with her husband Max and her children Mollie, Libbie, and Simon, all of whom must've changed considerably since she'd been gone, the youngest maybe having no memory at all of their mother.

Although the actual miles she cycled are questionable—it seems likely that Annie made most of the trip from France to San Francisco by boat—she no doubt rode through extreme temperatures and terrains, fending for herself in a way that was utterly unconventional for her time. She stepped away from the traditional roles for women, blurred gender lines by dressing in a man's riding suit and riding a "man's bicycle," and pioneered marketing for female athletes. She rode roughshod over the limitations that circumscribed the lives of women in order to seek fulfillment for her unconventional dreams. It's an amazing story of self-promotion, chutzpah, and unlikely athleticism, of a woman refusing to be defined solely as a wife and mother, and yet it was all but forgotten until Peter Zheutlin decided to research his great-grand-aunt's journey in the late 1990s.

Over the course of her adventures, she became a symbol for an entire generation of female cyclists and a pioneer in the struggle for equality. As Zheutlin puts it, "The hopes and aspirations of millions of women were riding on her handlebars." In her own account of her journey, Annie wrote, "I am a journalist and a New Woman—if that term means that I believe I can do anything any man can do."

In the months that followed, Annie moved her family to New York and had a brief stint as a feature writer for the *New York World*, Nellie Bly's newspaper, where her assignments were all about women working in new roles that had recently opened up to them. She sorted post on the New York mail train (once the exclusive preserve of men) and wrote about a women-only stock exchange off Wall Street. In 1897, almost two years to the day after the end of her bike trip, Annie and Max had a fourth child, Frieda.

It's not known if Annie ever rode a bicycle again, although I like to imagine she did. According to her family, she spoke often and with pride about her audacious adventure. Annie's granddaughter Mary said of her: "My grandmother had to be moving all the time." She died, of a stroke, in 1947, aged seventy-seven. There's no way of knowing how many women her "whirl around the world" inspired or empowered. But whenever I get on my bike I like to think of her, dressed in long skirts cycling along the Hudson river, or in her bloomers crossing the desert in Arizona.

Annie's story is also a reminder that for many women across the world, cycling is still a symbol of much-needed freedom. In Afghanistan, women who cycle have been labeled "infidels"; in Saudi Arabia they need a male chaperone to get around by bike. Many women continue to cycle in Iran, although some have been arrested for doing so. In Yemen, where a combination of fuel shortages and cultural norms has left many women unable to get around, photographer Bushra Al-Fusail launched a campaign in 2015 to encourage women to get out on their bikes and break the taboo. Her pictures of women cycling were met with dozens of furious comments online. "This can't be real, these images were photo-shopped," commented one Yemeni man. "Those are not women, they are men dressed as women," said another.

Cycling isn't just a way to get from one place to another, it can be a tool for personal and political power. Thanks to electric bikes and tandems, cycling can also be a game changer for those with physical barriers to walking, such as age or disability. The British charity Re-Cycle is providing women in various countries across Africa with recycled bikes as a safer and quicker mode of transport than walking. In LA, the Ovarian Psycos Bicycle Brigade are a cycling sisterhood for womxn of color, seeking to address the lack of diversity in the biking community. Getting on a bike is a way of opening up the world, of expanding your horizons far beyond what you could access on two feet.

There is something transformative about getting on a bike, or swimming, or screaming in a crazy exercise class, that frees you from any awareness or self-consciousness about your body or your physical limitations. When you combine that feeling of weightlessness with the freedom of travel it can have pretty memorable and magical results. Martha Gellhorn, a big fan of a skinny-dip in the nearest body of cold water, called this "a state of grace…when body and mind rejoice totally together. This occurs, as a divine surprise, in travel; this is why I will never finish traveling."

Say no to drugs when in control of moving vehicles, though. I learned that lesson the hard way.

CHAPTER FIVE
A Man's World?

Jeanne Baret and the women dressing as men to travel

When I got back from Shanghai, I had to face up to the reality of my new single life. But I didn't really want to. I'd decided to stay at my friend Josh's house for a while, but because I was traveling so much, I'd barely unpacked. It became little more than a crash pad in which to spend a few days between trips, do my washing, pack up, and leave again. Josh is a filmmaker and so was also away a lot, and he had only just moved in too, so for the most part we didn't even have any furniture.

I remember one night, heating up SpaghettiOs on toast—I never bothered having any food in the fridge because I was never there long enough to eat it—and then sitting on the floor to eat them because we didn't own a single chair. I had a flashback to my old life, where I'd cook multi-course Sunday lunches in my beautiful kitchen for a group of close friends. I felt sick when I thought of everything I'd given up, and for what? I cried myself to sleep that night, thinking about what an idiot I'd been to actively ruin my life and replace it with something much worse. This felt like rock bottom, but then every time I'd thought that before (when I told Sam I didn't want to be married any more, when I'd left our flat, when I broke up with Guy…) worse was yet to come. The next morning, after hours of sobbing in the fetal position, I caught the first flight out of London to Dubai for my next assignment—bringing new meaning to "the red eye."

As I was discovering more women who had found empowerment and freedom through traveling by themselves, one of the biggest things that struck me was how little has changed for female solo travelers today. When I was with Sam, people would express surprise and skepticism that I could want to travel without him. Now I was single, I'd regularly come up against the "Aren't you too old for another gap year?" attitude, or a mixture of pity and panic that at thirty-five I was deviating from the still-really-quite-expected track of wife and mother. Then there are the risks we associate with being a lone woman out in the world. Women are pre-programmed from an early age to think about threats and dangers and adapt our behavior accordingly. Anyone who has ever turned their keys into a makeshift mace while walking home alone at night or thought about who the safest person is to sit next to on the bus knows this to be true.

There is an intrinsic difference when navigating the world as a woman. There are things men do regularly—hitchhiking, walking freely at night, sleeping outdoors—that we tend to worry about women doing. While traveling alone, I have been advised by well-meaning people to stay indoors after dark, to wear baggy clothes, to cut my hair. But as I've got older and more confident, and come to recognize the spaces where men will hassle me, the less I care. I'm lucky that because of my white, cis, middle-class body, the world has generally been a welcoming and safe space for me, even without having to disguise myself as another gender.

But across the centuries, many women have pretended to be men in order to obtain more freedom while traveling and avoid unwanted attention. In 1768, twenty-six-year-old Jeanne Baret became the first woman to sail around the world. And she did it by dressing as a man. The story of how Jeanne became Jean is a fascinating and complicated one. And yet because no writing of her own survives—although she was almost certainly literate—we can only guess at her feelings and motivations.

What we do know is this: Jeanne was born on July 7, 1740, the daughter of farmers from France's Loire Valley. She probably became an "herb woman"—rural women, often midwives, who had exceptional knowledge of the medicinal properties of plants. At some point in 1760, she had an encounter that would change her life. While out collecting plants near her home, she met the renowned botanist Philibert Commerson. After the death of his wife she became his housekeeper. They were lovers by 1764; in the summer of that year, Jeanne was pregnant and Commerson was widely assumed to be the father.

They moved to Paris to study plants together and had a son who they gave to a foundling hospital (a not uncommon practice at the time). Commerson was invited on board a French ship called the *Etoile* to take part in an expedition led by explorer Louis-Antoine Bougainville, which hoped to redraw the known world. Commerson was invited to bring an assistant on board, but French law expressly prohibited women on ships. So at some point the couple came up with the idea of disguising Jeanne's gender.

I don't have any stories of adopting a male disguise on my travels, but I did used to get mistaken for a boy a lot when I was growing up. My dad had been desperate for a son, which meant we spent Saturdays watching Queen's Park Rangers, fly fishing, and playing chess. Although these shouldn't be solely male pursuits, I've realized as I've got older that not many women know how to fly fish (and it gives me a lot of "cool girl" points whenever I've told the men I've dated).

While I relished my tomboy status growing up (it's a rare photo that captures me in a dress, not dungarees), around puberty I started to be very self-conscious of the fact I didn't look like a girl. For some reason, I had decided to cut my hair extremely short (I was a big fan of the Gwyneth Paltrow movie *Sliding Doors*). But a pixie elfin crop on a beautiful Hollywood actress doesn't exactly translate to an awkward, undeveloped thirteen-year-old in Surbiton. I remember accompanying my dad to his squash club one night, only for a man in the bar to ask him how old his son was.

This was the 1990s, and Pamela Anderson and Wonderbra ads were everywhere. Which made it even worse that I was less "Hello Boys" and more "Hello, are you a boy?" To add insult to flat-chested injury, I remember lots of earnest talks at school about buying your first bra, starting your periods, and becoming "a woman" (said with dramatic emphasis). But none of that stuff was happening for me. My chest looked like two peas balancing on an ironing board and every new day brought fresh disappointment to the question everyone was asking: "Have you started yet?"

One of my teenage diary entries reads: "If there is a god, he must really f**king [yes, bad words asterixed out…I was that much of a goody-two-shoes] hate me for giving me such tiny tits." Wrestling with the metaphysical *and* my AA-cup in one heart-wrenching sentence. I never had a blossoming into womanhood, but as I grew out of my gangly teenage years, I did become more at peace with my body, which remains resolutely boyish.

There have been times when I've wondered whether being mistaken for a man while traveling could be an asset. The one place I did take some measures to suppress my femininity was India. When I was twenty-seven, I decided to quit my job and spend three months traveling through India and Nepal. Although I would've been too embarrassed to say it out loud, I was definitely trying to "find myself." As I approached thirty, I felt like I had achieved lots of the things I thought I was supposed to want—great friends, a sparkling career, a brilliant boyfriend, a foot on the property ladder. But I still felt like I was missing something. I'm not sure why I expected to find it in India, but it seemed like it might be worth a try.

Countless people warned me that I could expect to be hassled a lot traveling alone as a young woman with long blond hair, so I dyed it brown and cut it much shorter. I wore baggy clothes that covered me up and a gold band on my wedding finger. I spent the first few weeks being super-cautious about every little thing. Though, at first, India was such an assault on all my senses that being harassed by men seemed like the least of my worries. I strapped a weighty money belt of my possessions to me and only took it off to shower. I was convinced that every mouthful of food was going to make me sick. I didn't venture out alone after dark and during the day I was convinced that I was going to be mown down by a tuk tuk/car/motorbike/donkey every time I tried to cross the road.

Jeanne's first few days on board the ship were difficult, too. She would have been the only woman among more than three hundred men. She would have been unable to use the "heads" (holes cut into a projecting area of the deck at the front of the ship for sailors to relieve themselves), and senior officers and gentlemen shared a single commode. But on February 1, 1767, the *Etoile* set off with "Jean" aboard. The next port of call would be Rio de Janeiro.

Glynis Ridley, in her book *The Discovery of Jeanne Baret*, calls Jeanne "one of the boldest cross-dressers in history," which might well be true, but she certainly wasn't the most convincing. It wasn't long before members of the crew had their suspicions about the "little man" on the ship. François Vives, the ship's surgeon, wrote in his journal that "after a few weeks…a rumor, circulating among the men, concerning a girl in disguise on the ship. Everyone knew where to look." He recorded that Jeanne was made to leave Commerson's cabin and sleep in the lower decks in a hammock with the other servants. She slept with a fully loaded pistol. According to Vives, Jeanne insisted she wasn't a woman but an eunuch.

Jeanne battled seasickness, a skin condition caused by chafing from the linen bandages she used to bind her breasts, and the constant threat of detection. At every stop, she lugged heavy satchels and field bags full of instruments over unfamiliar terrain looking for new plant specimens. It was grueling physical work for any man or woman, and Commerson called her his "beast of burden." She endured periods of starvation and scurvy when provisions ran low.

In Rio, it was undoubtedly Jeanne, not Commerson, who brought to the Western world's attention the bright pink tropical vine that they would go on to name *bougainvillea*, after Bougainville, the commander of the ship. According to his diaries, Commerson was immobile and suffering from a recurring leg wound at the time. So it's thanks to Jeanne Baret that this splurge of color, which we so associate with warm climates, whether it's white-washed villas in Greece, or wedding decorations in Bali, has bloomed around the world.

Although there were ship-destroying storms, hurricane-force winds, and grim hazing rituals such as "Crossing the Line" (when the ship crossed the Equator, everyone stripped and was dunked in "the Pool" of seawater...and worse), there were wonders, too. Jeanne would've watched barnacle-encrusted humpback whales, elephant seals, penguins, twenty-foot manta rays, and dolphins. Commerson modestly named a small, blunt-headed species of dolphin after himself, a bit of a *faux pas* in taxonomy, apparently. Jeanne would've seen the glow of bioluminescence, gazed at the Great Barrier Reef (back then no doubt a pristine multicolored coral kingdom, rather than the rather desiccated pale version we see today), glimpsed her first solar eclipse and spent nights beneath a wholly unfamiliar pattern of stars in the southern hemisphere.

While looking for Indonesia, the ship bumped into a landmass that turned out to be Tahiti. It is at this point in Bougainville's journals that he first made reference to Jeanne. "For some time there had been a report in both ships, that the servant of M. de Commerson, named Baret, was a woman. His shape, voice, beardless chin, and scrupulous attention of not changing his linen, or making the natural discharges in the presence of any one, besides several other signs, had given rise to, and kept up this suspicion."

Bougainville went on to describe the ship landing at Tahiti and the local men clocking Jeanne's true identity. "After that period it was difficult to prevent the sailors from alarming her modesty," reported Bougainville. While Jeanne, who he described as "neither pretty nor ugly" with a "faced bathed in tears, owned to me that she was a woman." "She will be the first woman that ever made it, and I must do her the justice to affirm that she has always behaved on board with the most scrupulous modesty."

Toward the end of the trip, any doubts about Jeanne's gender were permanently put to rest in a traumatic way. Vives wrote that, while washing her clothes in New Ireland, Jeanne was "examined," the implication being that she was assaulted and possibly even raped. Soon after, the ships docked at Île de France, today known as Mauritius, and Jeanne and Commerson stayed there, happily botanizing, until his death in 1773. Although Commerson had written Jeanne into his will before they left Paris, she would have to get back to France to claim it and she had no money. We know she found work as a barmaid in a tavern in the capital of Port Louis as Mauritian records show that she was fined for selling alcohol on the Sabbath in 1773. Here she met a soldier, Jean Dubernat, and they married in 1774 and sailed back to France.

When Jeanne touched French soil in 1775, she had made history and yet there was no welcoming committee or celebration. No statues were erected in her honor. It's likely that no one but her knew what she'd achieved. Only a single image of Jeanne survives—an intriguing watercolor dating from 1816 of a woman in loose-fitting striped linen trousers and a wide-cut tunic. Nine years after returning home, she finally got some recognition and received a special pension from the Ministry of the Marine "bestowed upon this extraordinary woman," perhaps making her the first woman recognized by any state for her contribution to science.

Prince Nassau-Siegen, the flamboyant royal on board who was

mistaken for a woman because of his blond wig and makeup, wrote in his journal about Jeanne: "I want to give her all the credit for her bravery, a far cry from the gentle pastimes afforded her sex. She dared confront the stress, the dangers, and everything that happened that one could realistically expect on such a voyage. Her adventure should, I think, be included in a history of famous women." But of course, it wasn't.

Jeanne had helped Commerson collect more than six thousand specimens, but not a single plant was named after her, while more than seventy species today bear the name *Commersonii*. In 2012 she finally got the recognition she deserved, when a new species of Solanum vine was discovered in southern Ecuador and northern Peru by Eric Tepe and named *Solanum baretiae* in Jeanne Baret's honor.

Although Jeanne is one of the first we know about, dressing as a man in order to travel became a not uncommon tactic for women, for practical as well as socially subversive reasons. The French writer Amantine Dupin, whose pen name was George Sand, frequently donned male clothing to walk the streets of Paris in the 1830s. Scottish teenager Isobel Gunn was only fifteen when she disguised herself as a male laborer and traveled to the Canadian wilderness in 1806 to work as a fur trapper. Her gender was apparently discovered only after she gave birth at a trading post.

But even that straight-out-of-a-soap-opera story doesn't seem shocking compared with Isabelle Eberhardt's tumultuous life. Born a biological woman in Geneva in 1877, the illegitimate lovechild of Russian anarchists (though there were also rumors she was the love child of the French poet Rimbaud), while still a teenager she published short stories under the pseudonym Nicolas Podolinsky and began wearing male clothing to move around more freely.

In 1897 she moved to Algeria where she dressed, wrote, and lived as a man. She began calling herself Si Mahmoud Saadi. When asked why she simply said, "It is impossible for me to do otherwise." Perhaps she would've identified as transgender or nonbinary today. Eberhardt became fluent in Arabic and converted to Islam. But she wasn't exactly devout.

According to a friend, Eberhardt "drank more than a Legionnaire, smoked more kief than a hashish addict, and made love for the love of making love"—with both men and women. She spent inherited money recklessly—including throwing large amounts of it out of the window—and chose to live in destitution.

Eberhardt's diaries are full of the intense bliss and heady excitement of traveling. "Oh that extraordinary feeling of intoxication I had tonight, in the peaceful shadows of the great al-Jadid Mosque during the icha prayer! I feel I am coming back to life again." Her joy at being "far from society, far from civilization" is as palpable as the heat and the flies and the dead camels she describes. "I am by myself, on Muslim soil, out in the desert, free and in the best of circumstances." She fought the police in Algerian uprisings, survived an assassination attempt with a saber that nearly severed her arm, and died during a freak flash flood in the desert at just twenty-seven.

I've no doubt that my experience of traveling through India would have been very different had I been a man. Although this was a year before the gang rape and murder on a bus in Delhi shocked the world, public transport in India had long been notoriously unsafe for women. A 2018 Thomson Reuters poll listed India as the most dangerous country for women with an estimated four rapes taking place every hour. On one bus journey, a man wouldn't stop talking to me. Which would have been fine if he didn't have his hands inside his trousers the whole time. In true British fashion, I pretended to ignore him and moved away, until an old woman saw what was happening and started beating him with a stick and berating him with words I couldn't understand but no doubt would have wholeheartedly agreed with.

Overfriendly men (staring, taking photos, and worse) became an all-too-common occurrence for me. I went on an overnight camel trek through the Thar desert outside Jaisalmer where, as predicted by my guide book, the two men leading it offered to give me massages. Er, thanks but no thanks. They soon backed off without bothering me further. But the scariest moment happened when I arrived in Jodhpur, the Blue City, just before dawn.

I had got off my overnight sleeper train and haggled for a tuk tuk to take me to a hostel. We arrived at a rundown, crumbling six-story building where an old man was dozing at reception. He said he had a room, and we walked up what felt like endless flights of stairs. I remember he had watery eyes and was unshaven. He opened the door to the room, gestured for me to go in, and then followed me inside while I looked at the bathroom (experience had taught me in India to *always* check out the bathroom). But I already knew that I didn't want to stay here—there was a weird feeling about the place.

I said I'd go and get the rest of my luggage and planned to do a runner, but when I tried to leave the man blocked by way and stood too close. He smelled like stale alcohol. He made a very awkward and almost half-hearted attempt to push me against the wall. Thankfully I was not only taller but stronger (or at least less drunk) and was able to push him off, race down the stairs and get straight into the tuk tuk that was still waiting outside with my luggage.

I remember feeling strangely calm as we drove to another hostel. But when the tired teenage boy at reception told me, with a head wobble I had come to know and love in India, that I could wait until someone "probably" checked out that afternoon, I promptly burst into tears. He kindly made me a cup of hot, sweet chai and then I got onto a table in what looked like a store cupboard and slept for six hours. When I woke up, there were chickens and puppies and children around me and a woman peeling potatoes. I put the whole incident out of my mind. I didn't even bother mentioning it in my emails back home. I knew I was lucky it hadn't been worse.

So I completely understand why, during the 1970s, Sarah Hobson, then twenty-three, decided to disguise herself as a man named John while traveling through Iran. In her book, *Through Persia in Disguise*, Sarah describes dressing in drab, shapeless clothes and shoes two sizes too big, adopting male mannerisms and covering her "bulges" with a chest girdle and notebooks in her breast pocket. She had to control her bladder and only go to the open toilets at night. She kept her nails trimmed and let her eyebrows grow out but was almost given away when someone spotted her pierced ear lobes. "I'm part Scottish…it's our clan you know. All the men have their ears pierced," she stuttered. But dressing as a man was not enough to prevent sexual harassment from both women and men—even though homosexuality is illegal in Iran.

However, there are some benefits to being a boy. All that effort seemed to be worth it when it allowed Sarah to gain access to and eavesdrop in spaces where no women are allowed, from the deep theological debate in a *madrasa* (a male-only Islamic college) to the banter in the teahouse, which centered on the number of kebabs a man could eat without being sick. Like an undercover cop who goes too deep, she occasionally forgot her true gender and sexuality. "Sometimes I felt the disguise had taken over, that I really was a boy," she wrote. "Quite unconsciously I would size up the shape of a woman's legs, or speculate, with the other boys, what important jobs we would soon secure as men."

Occasionally she had to pretend—like Jeanne Baret—to be a eunuch. She was almost arrested as a Russian spy when the female photo in her passport didn't match her face. But for the majority of her time in Iran, she was welcomed as a man. "It seemed that most people I met accepted me as a boy," she wrote. "Perhaps I had grown more masculine, but probably they took me at face-value and put down any doubts to the difference of culture and habits." Those who she spent a sustained amount of time with suspected her secret, or even acknowledged her gender fluidity, calling her "Miss John."

▪——➤

I don't want to portray India as a bad destination for solo female travelers. The uncomfortable moments were far outweighed by the positive interactions I had with strangers there. India was everything I had imagined and more—more colorful, more chaotic, more challenging. Although I didn't quite "find myself," I had an eye-opening time trying. I visited an ashram where the mode of prayer was hugs. I rode pillion on the back of a motorbike speeding through Calcutta, missing cows by inches. I spent my twenty-seventh birthday in the foothills of the Himalayas with a monk wearing a Manchester United shirt. I felt like all the air had been sucked out of the atmosphere when I walked through a dark arch and saw the Taj Mahal for the first time—one of the few instances when something that's on every "must-see" list is actually a must-see.

Looking back, there's no doubt that surviving India alone made me a better traveler and a more confident person—which is another brilliant thing about solo travel. The more you push yourself out of your comfort zone, the more your confidence in yourself and what you're capable of swells. Now whenever I find myself in a sticky situation, I remind myself, "I went to India for three months and survived—I can handle this."

Louis Bourgainville said of Jeanne Baret, "She will be the only one of her sex [to travel around the world] and I admire her determination... her example will hardly be contagious." But he was wrong. Women were only just getting started.

As for me, I couldn't stop traveling. It had long been an easy way to escape from the doubts I would occasionally have about my relationship with Sam, and now it was an escape from the guilt I felt about leaving my marriage, about how much I'd hurt Sam and his whole family, the mess I'd made of everything. My life in London was in tatters and getting on a plane—it didn't matter where to—provided brief respite from the pain. I knew something had to change and I had to put down some roots on my own (or that I should at least unpack a bit) but I couldn't think about the future, only my next trip.

STAYING SAFE WHEN
YOU TRAVEL SOLO

While there's a lot to be said for not thinking you're going to be groped by every person who brushes past you or picturing your plane plummeting to the ground with every jolt of turbulence, it is good to take some measures to make sure you feel secure. Mainly because when things go wrong, it's really annoying when you don't have someone else to blame it on.

1. **Practice daylight savings.**
If you can, try to arrive in a new place in the daytime. It means you'll have far more of a sense of where you are and what's around you. It also means you have time to change your accommodation if it's not what you expected. Plus everything seems scarier at night; it's just a fact.

2. **Don't be polite.**
If you're being made to feel uncomfortable, don't be polite about it, don't be a people-pleaser—speak up. I often think about all the potentially life-threatening situations we put ourselves in because we want to be polite, whether it's accepting a lift from

someone who drives too fast or being too embarrassed to ask for help when we need it. Fuck polite. If it doesn't feel right, do something or say something. And loudly, if possible.

3. Be aware.

You don't have to have your guard up the entire time you're away, just practice all the things you would normally do at home. Don't wear headphones so you can't hear what's around you. Don't wave your phone around. Don't leave things unattended, even for a moment.

4. Lock your door.

From the inside too. It's a good habit to get into. I'd love to say I also make myself aware of where the fire exits are where I'm staying. Let's say I will from now on.

5. Have a back-up credit card.

I was in Mexico when my credit card—my only source of funds—was sucked up into an ATM in a tiny border town. Cue many long exasperating phone calls to my bank back home. So take a spare card if you can, and don't keep all your cash and cards in one place—spread the wealth. Keep a bit in the safe in your room (if there is one), a bit secured in your suitcase, a bit in your wallet, some in your bag, some in your bra. I have a sports bra from Lululemon with secret pockets in it, which is handy for keeping some cash and a card in (which for me is preferable to a bulky money belt). I also try not to get too much cash out at once, although if you're traveling somewhere with limited ATMs (Cuba, the rainforest) this obviously isn't practical.

6. Tell someone where you're going.

It might be letting the hotel receptionist know your plans for the evening. It can be WhatsApping your friend back home. If you're somewhere where you feel at all concerned for your safety, you might want to consider a GPS tracker app. TripWhistle will log your location and send messages to family or local emergency services if you're in trouble. And on that note, it's good practice to find out the number for emergency services where you are and save it in your phone. You don't want to wait until you really need it to discover it's not 999 the world over.

7.   Try to look like you know where you're going.

Pickpockets are known to target those who look vulnerable or distracted, which is why tourists are easy bait. That and the massive backpacks. Although it's difficult in unfamiliar territory, I try to always walk with an air of confidence, not clutching a map or staring rigidly at the blue dot on my phone.

8.   Learn some lingo.

This isn't just about staying safe; it can enhance your whole experience of a country. I'd love to say that I'm a natural linguist but sadly I am impressively hopeless. I once secretly took three months of Italian classes so that I could impress Sam when we arrived in Tuscany and could still barely order a coffee. But memorizing (or writing down in your phone) a couple of key words and phrases can make all the difference. Along with the basics you'd expect, "Help me," "No," and "Stop" are useful to have in your arsenal, even if you only use them when you're panicking because you've missed your stop on the bus.

CHAPTER SIX
Fight or Flight?

Elspeth Beard and using travel to heal trauma

I wasn't alone in using travel as the answer to my whole life falling apart. Often it's a big life change—and a traumatic one at that—which becomes the catalyst for a solo adventure. Getting divorced, widowed, dumped, or even a big health scare can all precipitate the need for a trip of one's own. For Elspeth Beard, the impetus for her becoming the first British woman to go around the world on a motorbike in 1983 was something that nearly all of us have experienced—brutal heartbreak. While still a twenty-four-year-old architecture student, she was dumped by Alex, "the guy I thought was going to be the love of my life," via letter, a week before Valentine's Day.

"Desperately unhappy and lonely, I looked for distractions," she wrote in her memoir *Lone Rider*. "To get away from Alex, maybe to prove something to him, I decided I'd now definitely go away." She admitted that "riding a bike around the world was probably an extreme reaction to heartbreak," but she knew that a two-week beach holiday wasn't going to cut it.

With just a few bags, a small amount of savings, "and yearnings for my ex-boyfriend in my heart," she shipped her bike to New York and set off all alone. This was long before mobile phones, Internet, and email. Elspeth didn't have proper maps, let alone sat nav. In many places, she had to make do with "photocopied scraps." Although she wrote to many magazines, potential sponsors, and bike manufacturers for advice and publicity, no one was interested in her story. Biking was even more of a male-only club back then than it is now.

At first, memories of her ex-boyfriend still haunted her. "Two hundred and seventy miles of desperately dull highways…inevitably, my thoughts turned to Alex…I wondered would I even be here if he hadn't dumped me." But Elspeth quickly discovered how much she enjoyed being by herself. "Those moments when all thoughts of the past and future slipped away and I existed entirely in the present, the miles rolling past beneath the wheels of my big BMW, the morning light clear and golden, throwing shadow bands across the road as I carved my way around the world…I felt a great sense of freedom. I could go where I wanted, when I wanted, stop when it suited me." She experienced one of the best things about going away on your own—solo travel means zero compromise. It's rare to find a travel buddy who shares your niche enthusiasms, body clock, appetites, budget, and/or desire to visit a pencil museum. But when you're solo you can do exactly as you please.

Women seeking redemption or reinvention through travel is nothing new. *Eat, Pray, Love*, Elizabeth Gilbert's year of snacking and self-discovery, was famously motivated by a messy divorce, a bad break-up, and a total life crisis. You only have to witness the queue for L'Antica Pizzeria da Michele—the pizza restaurant in Naples where Gilbert has her "eat" epiphany—to see the impact this multimillion-selling memoir, published in 2006, continues to have on generations of women (me, unashamedly, included). We all want to believe that if we're going through a bad time, one trip (ideally with a hot new boyfriend discovered along the way) can turn it all around.

Similarly for Cheryl Strayed, a 1,100-mile hike became the antidote to being "unmoored by sorrow." In 1995 aged twenty-six, she made the snap decision to walk along the Pacific Crest Trail from Mojave, California, to Washington, alone, despite having never gone camping before and unable to even lift her backpack. She had just lost her mother to cancer, just got divorced, and was battling depression, eating disorders, and a drug habit. In her 2012 book, *Wild: A Journey from Lost to Found* (another bestselling-memoir-turned-Hollywood movie), Strayed describes outwalking her suffering and finding herself, one blistered step at a time.

The flipside of carefree independence is loneliness though, and for Elspeth Beard this soon took its toll. Although riding the bike for hours every day was physically exhausting, the mental challenges of doing it alone were harder. "Never knowing where I might find water or fuel… always on edge, unable to relax, anxious and tense." Even when she met other travelers in hostels, her experiences on a bike and by herself meant that her stories were very different to theirs. When her parents came out to visit her, she felt miles away from their luxury hotel and organized coach tours. "I didn't fit in with my parents, or with other travelers, or with the locals," she wrote. "Ultimately, I didn't belong anywhere; a very lonely place to be."

Elspeth spent months riding through countries where she didn't speak the language, couldn't understand the road signs and where there were often no roads—just dusty tracks, mudslides or sand. In places where the culture discriminated against women she had to keep her helmet on in public so that everyone presumed she was a man. In Arizona, she was chased by a biker gang. In Australia, she was shocked to find that she was having a miscarriage in a hostel dorm room (she hadn't even realized she was pregnant). In India, a man pushed his daughter in the way of her bike in order to claim "compensation" of only a few rupees for her injuries. Elspeth battled hepatitis, giardiasis, and dysentery. All her valuables—passport, money, bike keys—were stolen in Singapore. In

Thailand, she hit and killed a dog, wrecking her bike. The family who welcomed her in to stay and recover served her many delicious meaty curries. I think you can guess where that story was going.

Accompanying Elspeth through it all was her "old girl," aka her bike. "We were like a dog and its owner, although I wasn't quite sure which of us was the master." She had to repair every breakdown and fault herself, often without proper tools or parts. At one point, she used a heated-up Malaysian coin instead of a soldering iron. Her bike broke down so often, it began to feel like more of a liability and less of a friend. After more than 35,000 miles, Elspeth was riding with no front or rear lights, no back brake, and a lethal front brake. She admits that, "Until we reached Turkey, I wasn't even certain we'd get home." Over the course of her trip, she fell in love with two very different men, each with unexpected and long-lasting repercussions. For all the pitfalls, it was exciting, eye-opening, and the defining experience of her life.

When she finally rode her bike back through London at 5:00 a.m., after more than two years on the road, she wrote that her home city felt "still, silent, and entirely alien." In many ways, her biggest challenges were still ahead of her. Even her biking friends found it hard to relate to what she'd been through. Elspeth experienced the big difficulty with solo travel—that when you feel so changed, it can be hard to slot back into your old life. Instead, she retreated into herself and felt more isolated in London than she did "in the middle of the desert or on the side of a mountain."

Even before my life imploded, I'd often find it difficult to come home after a solo trip. I'd spend a long time looking forward to seeing my friends again only to find it a jarring experience when we did finally meet up. Things I'd fantasized about—peanut butter on toast, my own bed—weren't as great as I remembered. I have never felt as alienated as Elspeth Beard did after almost two years away from home, but even after a shorter trip I think you can feel different and out of step with your old self, and it takes at least a few days to acclimatize.

After her adventure, Elspeth Beard faced a series of difficult and harsh life events and was driven to the brink of suicide while attempting to convert an old water tower into her home while single-handedly looking after a six-month old baby. "It was the toughest period of my life, harder than anything on my global bike trip," she wrote in *Lone Rider*.

Thirty years on, Elspeth thought her round-the-world ride had been "forgotten by almost everyone" but in 2008 an article about her trip was published in an online biking magazine and her story spread. She received interest from a Hollywood producer who wanted to make a movie about her. In 2017, she published her memoir and, when I first heard about her story in an online article about female bikers, something really resonated with me. Whether it was the heartbreak I could identify with or the fact I'd dreamed for a long time of getting my motorbike license, I really wanted to meet her.

Elspeth still lives in her award-winning converted water tower in Godalming, Surrey, so when I found myself in the United Kingdom between trips, I arranged to pay her a visit. Now aged sixty-one, with cropped dark hair and dressed in all black, she was just as striking as the twenty-three-year-old biker in the black-and-white photos I'd seen of her. I felt a bit nervous to meet her. In the book, she comes across as so no-nonsense and impressive and while reading it I was intimidated by her coolness. Luckily, in person she was friendly and laughed a lot, immediately putting me at ease.

"I do still have the bike and I still ride her!" she said. "I hadn't run her for fifteen years, but when I put in petrol and a new battery she started straight away! She's tired though; she's done her time. I've still got my leather jacket and my helmet, too."

I told Elspeth about my own story and the too-often-forgotten female explorers I was researching and she seemed genuinely enthused. "I'm staggered by how many women did these amazing trips in the 1920s and even before that," she said. "I think these stories have remained untold because women do these kinds of trips for themselves, whereas men do it to be like 'Look what I've done!' When I think of the negativity I faced and everyone telling me I couldn't do it, and it must've been so much harder for these women before me. They must've been so focused and determined to ignore everyone else and do what they wanted."

After her round-the-world trip by bike, Elspeth's globe-trotting was far from over. "In the nineties I got my pilot's license and in 1994 I took my son, Tom, to Australia and hired a plane and flew around western Australia," she said. "Then I got a camper van and drove that for around 30,000 miles around Australia. In 1996, I flew out to South Africa and drove through Namibia, Botswana, and back into South Africa. I went to Tibet in 1998 by bike, and Peru, Bolivia, and Chile."

She said that writing the book made her realize how much her first trip had shaped her as a person. "I doubt I would've had the self-belief or courage that I was going to overcome all the obstacles that have come my way if I hadn't made the trip. But at the time I had no idea about the impact it had on me. All the interest in my bike trip has happened in the last five or six years and so it's all come back around. Something that changed my life so much in my twenties is now changing my life all over again, which is weird."

For Elspeth, travel now is almost unrecognizable to the way things were in the 1980s when she went around the world. "It really did feel like going out into the unknown," she explained. "I honestly didn't know if I would ever make it back."

I asked her if she thinks something has been lost by that, in our ability to be connected at all times now, even when traveling. "Absolutely!" she said. "Although travel was harder back then, I wasn't wedded to my phone and blogging every night. I lived every minute of every day. It was survival."

The biking world has changed too. "Back in the eighties, my trip was too much for these macho bike magazine editors to get their heads around so it was easier for them to ignore it. All my biking gear was made for men because women's sizes didn't exist. My boots were stuffed with newspaper at the end because that was the smallest size I could buy," she said. "Now bike companies have woken up to the fact that the other 50 percent of the population can ride bikes, too. I've had some great letters from women who have read my book, and it's inspired them to do something similar. A guy wrote to me to say that he was really into bikes and his partner wasn't and he'd been trying to get her to go biking with him for years and she wouldn't. Apparently she then read my book and within a month she'd taken her test and she was currently riding her motorbike around Europe without him!"

Elspeth now guides other motorcyclists—many of them women—on their own adventures. Last year she led an all-women's ride through the mountains of northern Pakistan. "Apparently lots of women come on the trips because they want to meet me, which I find really bizarre," she laughed. "Last year, I was in Spain, Portugal, Germany, Poland. In Lisbon we did this amazing night ride through the streets with nearly 700 motorbikes at 1:00 a.m."

Elspeth believes that in life you have to be prepared to take risks, grab opportunities, and not be afraid. "Three years ago I was driving through Italy and I saw this old ruined building with a church attached and a "For Sale" sign on it, so I thought, *I'm going to buy that*, and three weeks later it was mine," she said. "I feel like I've crammed so much into my life it's ridiculous, but I have no regrets at all. Not every aspect is perfect, but I've done exactly what I wanted and I really have lived."

I asked Elspeth about Alex, the man who dumped her and inspired her trip in the first place. She put her hands over her face and laughed. "I think he was quite taken aback when he read the book and realized how hurt I'd been. Now I look back and I'm so glad he broke up with me. It made me realize that if anything bad happens in life—and at the time you think it's so dreadful—something positive always comes out of it. Years of experience have taught me that. You just have to hang on and wait for it to get good."

I wanted to know if there was a moment from her trip that most often comes back to her and she answered without hesitation. "When I fell in love with Robert riding through Kashmir," she said, tears instantly welling up. "That was…yeah." She began crying and couldn't continue. I nearly cried too. For both of us, the memory of heartbreak was clearly still raw—in Elspeth's case, even almost forty years later.

Although deep down I knew that a plane ticket wasn't a quick fix for my problems, I have long believed that getting away can provide some much-needed perspective. For me, a trip, particularly a solo one, has often proved cathartic. When you're spending time alone and in unfamiliar surroundings, things pop into your mind that surprise you. Although expecting yourself to come back enlightened from a trip is probably as successful as anxiously ordering yourself to "JUST RELAX," you may find that something in your mind has changed. And sometimes that changes everything. That's why I always carried the hope that this next adventure might be the one where I'd have a transcendent moment of clarity and figure everything out. But Satan in *Paradise Lost* has this to say about running away from our demons: "Which way I fly is Hell; myself am Hell." Or, more simply put: "Wherever you go, there you are."

My life felt less chaotic now. Three months had passed since I had left the safety of my flat I shared with Sam and gone to Israel. And then China. And then Dubai, Paris, the Highlands, the Maldives. I was operating on an "Anywhere But Here" policy. But in place of the fizzing whirl of change and transition I felt deeply lonely. Guy had respected my wishes and stopped contacting me. I knew I had to try to file that painful experience away as a fun rebound that had cheered me up at a time when I'd needed it most. I toyed with signing up for online dating but having been in a relationship for twelve years, the world of swiping right was a strange one, and not just because I once matched with a man who was riding a horse half naked in his profile picture, Putin-style. Or

"Martin XXL" who wanted me to know in his bio that he had "a very big manhood." I screen-grabbed (scream-grabbed?) terrible dating profiles and sent them to my friends to cackle over. But behind the laughter was a slowly rising panic that I'd never meet anyone and would die alone.

In a strange way, going away by myself somehow eased my loneliness for a little while. One of the ironic things about solo travel is that when you're by yourself, you're actually far more likely to meet new people. When you're traveling as a couple, no one will talk to you, even if you're not face-sucking all over the place, and that's because everyone assumes that couples want alone time. If you're traveling with even one friend, let alone a whole group, it can be intimidating for others to approach you. But when you're solo, everyone—from the chatty waiter to the bored couple—wants to talk to you.

The problem was that my solo trips, fun as they were, were only intensifying my isolation whenever I came home. Friends stopped inviting me to things because they presumed I'd be away—by this point, I'd missed a lot of birthdays, christenings, and even a few weddings because of work trips. When I did reach out to them, people expressed surprise that I was even in London at all because, according to my Instagram, I was living my best life in an infinity pool in the Bahamas. The reality was more crying for hours in my bed in New Cross.

I felt restless when I got home, and homesick when I was away. I was reminded of a line from Tony Kushner's play *Angels in America*—"it's the price of rootlessness. Motion sickness. The only cure: to keep moving." I made tentative plans to leave London and live abroad long-term, perhaps back to California. I had always thought of myself as a nomad—someone who could feel at home anywhere—but I was starting to feel at home nowhere. I realized that my travels had tipped from rewarding into running away. There are times when a trip is just the perspective and distraction you need, but this was something more. I knew I wasn't just trying to get away from life in London, I was trying to get away from myself.

HOW TO PACK FOR A SOLO TRIP

Packing is an art form, and over the years I have learned some tricks. There's nothing worse than overpacking—every time you pick up your suitcase you're reminded of the weight of your indecision. I once took a wheely suitcase to Glastonbury, rolling it through the mud while people pointed and laughed, so I should know. But equally annoying is when you woefully underpack. No one wants to go on holiday with that carefree person who bought a bag the size of an envelope and then spends all week borrowing your socks and reading your books.

In her 1979 essay collection, *The White Album*, Joan Didion revealed that she kept a packing list taped to the inside of her wardrobe door for when she had to leave on an assignment with no notice. Because it is Joan Didion, this list is at once impossibly stylish (mohair throw), ridiculously cool (cigarettes, bourbon) and effortlessly practical (nightgown, robe, slippers). I've always found mohair itchy, I don't smoke, and I hope I might one day be the kind of woman who owns a nightgown, but here is how I like to pack…

1. Take out a pair of shoes.
Just like Coco Chanel said to remove one accessory before you

leave the house, you need to remove a pair of shoes before you zip up your bag. Ideally that pair of shoes you don't even really wear at home because they start to hurt before you've even ordered the Uber. You need one comfy pair of shoes (probably trainers, which you should wear to the airport) and one slightly smarter pair which are also comfy but look a bit posher. That's all. Don't become the person who misses out on that epic walk to a waterfall because their shoes were rubbing. Unless you are doing something that requires specialist footwear (hiking Macchu Picchu, learning to tango in Buenos Aires, running the Paris marathon), you really don't need more than a couple of pairs of shoes. Joan backs me up on this two pairs of shoes rule, by the way.

2.   While we're talking shoes…

I once had a professional organizer clear out my wardrobe for a magazine article, and one of the things she told me has never left me. Your shoes have been treading on who knows what on the pavement, so ideally you don't want them touching your clothes. The organizer packed all her shoes in tote bags or dust bags. I try to remember to do this, but when you're packing for a holiday and the check-in desk closes in thirty minutes and you're still in your underwear in your bedroom, you won't be thinking about bagging up your shoes. You'll make the flight but now you'll spend it torturing yourself thinking about the dog poo touching your favorite jumper. You're welcome!

3.   On the subject of jumpers…

People who know about fashion will call this a lightweight knit. Joan calls hers a "pullover sweater." I prefer to call a hoodie a hoodie. It sounds obvious but many is the time I've checked the weather forecast in a destination and got over-excited and so brought not a single item of warm clothing. And then spent the evenings going slowly hypothermic but saying, "I'm fine!" and trying to style the restaurant table cloth as a shawl. Jumpers are very bulky and boring (there's no getting around this) but some kind of cozy cover-up is essential, even if you're going somewhere tropical. Planes can blow hot and cold like a moody lover. Even balmy nights can get unexpectedly chilly. If you're visiting religious sites you may need to cover your shoulders. I know I sound like a nana, but please bring at least one jumper.

4.   Be pouch perfect.

When I discovered the pouches trick, it changed my packing forever. By dividing your most-used things between different small bags (makeup bags work perfectly) you're never tipping your entire suitcase out on the floor while you desperately look for an adapter. I like to put my electricals in one pouch (chargers, adapters, cables, etc.). Sunscreen in another. (All right Baz Luhrmann, we get it.) Underwear is generally considered a good idea, and it's helpful to give it its own VIP area. Swimwear gets its own pouch for me because I have an inordinate amount of bikinis, and I will not apologize for this. A medicine kit might be another one (see page 110). You get the gist. This trick is especially handy if you're living out of a rucksack for a while and don't want to send yourself mad. Tote bags can work in a similar fashion and are also good for storing your dirty clothes once you've worn them.

5.   Pack one really fierce outfit.

Preferably a dress that folds up really small. Although I'm against the idea of a holiday wardrobe (not least for sustainability reasons), when you're traveling you want to feel like your best self, so only pack your favorite clothes. Don't pack that shirt that's missing a button or the shorts with a period stain on the crotch that you can't throw away because they're the best denim shorts you've ever owned and their like will never be seen again. But pack one dress (you really only need one) that makes you feel like you'd want to have sex with yourself if you looked in the mirror too long. Even if you're staying in a cottage in the middle of nowhere and so only end up wearing said dress to a weird pub that smells funny, who cares?

6.   Don't pack a full case.

Lots of people will tell you tricks about rolling not folding to squeeze in the maximum clothes possible. This is great until you have to unpack and pack again. Leave enough room in your case that you don't need an architectural drawing and a few hours to get everything to fit back in again. And maybe even give yourself some space to buy a souvenir or fit a plastic bag of soggy swim-wear because you went for one last dip in the ocean before your flight. (Always go for one last dip in the ocean before anything. It's always worth it.)

7.   Try to unpack.

Even if you're only staying somewhere a single night, unpack what you'll need, if you can. It provides a certain sense of sat-isfaction and security and somehow makes you feel a little bit more settled and grounded. Consider it a form of self-care, and it really only takes a few minutes.

8.   Bring comfy trousers.

You know those high-waisted jeans that look killer and are so tight they kind of give you a camel toe but you get away with it because they look awesome? Leave them at home. Traveling can be tiring, exhausting, stressful, and not temperature regulated.

It can also involve eating a lot. Some loose-fitting trousers are your best friend. A pair of Marks and Spencer cashmere tracksuit bottoms changed my whole traveling experience and maybe also my life. Perhaps you're even the kind of person who can rock a slouchy bottom with a heel and make it work from day to night (alas, I am not that person).

9.   Don't forget an eye mask.

When the bright orange interior plus fluorescent overhead lights of a budget airline are giving you a migraine, you need an eye mask. When the Airbnb you've rented is "warehouse style" but you discover this means no blinds, you need an eye mask. When you're sat next to someone annoying on the plane so you need to pretend you're sleeping...you get the gist. The best I've found is a Tempur-Pedic one (like the mattress) that molds to your face and makes you feel like you've strapped on a little pillow. Along with silicone ear plugs—it's my secret weapon to sleeping any time, any place. It costs thirty pounds though, so it's very annoying when you leave it on the plane, as I have done too many times.

10. Pack a medicine bag.

Although you don't need to be the person who brings water sterilizing tablets for a weekend camping in Cornwall, it is a really good idea to have a small pouch with a few of the non-fun kind of drugs in it. I leave mine zipped in a compartment of my suitcase, so I never have to unpack and repack it. Everything in it probably expired in 2005, but I still carry them because you never know. Hopefully you'll never have to use it, but if and when you do you'll be so overcome with gratitude you might just do a dance to the gods of ibuprofen. My very basic medicine kit contains rehydration salts (not just for when you're weeing out your bum, also great on a hangover), Theraflu (again, not just for colds but a miraculous hangover cure), acetaminophen, a couple of bandages, and a tampon. If you've ever found yourself searching for the Spanish word for tampon for the hot pharmacy guy in Bogotá and ending up doing a mime to demonstrate what you mean, you'll appreciate why this is handy. It should be said that this is the basic package and if you're going to a malarial swamp in the middle of nowhere you'll need to soup this up a bit (a lot)—unless you're a Bear Grylls type who can spin a mosquito net out of a panty liner. But trust me on the Theraflu.

CHAPTER SEVEN
The Traveling Cure

Gertrude Bell and the ladies prescribed a "change of scene"

I knew I had to do something about my sense of restlessness and disconnection. Perhaps I would still move abroad. But in the meantime, I needed to at least to try to settle a little into single life in London, to accept that this was where I had landed, even if it was only going to be temporary. So I finally unpacked my bin bags and boxes. I bought a plant. I made an effort to start cooking again and going to farmers' markets and doing a regular yoga class and every other middle-class cliché. Even if I wasn't quite ready to stop jetting off at every opportunity, at least I knew where all my belongings were where I came back. Traveling had served me well as a cure for heartbreak, and I wasn't able to stop yet.

I discovered that using solo trips as a panacea had a precedent. Back in the nineteenth century, many wealthy society women were using "doctor's orders" as an excuse to skip to the country and sack off a stifling and genteel life indoors. Far from the image we have of weak "angels in the hearth" who constantly needed the smelling salts passed to them, it turns out that a few Victorian ladies were bravely bossing it all over the world and enduring heat and hardship in order to feel free.

When Isabella Bird's doctor prescribed a "change of air" for her life-long illnesses and ailments, he probably didn't imagine that she'd end up trekking on horseback through the Wild West or climbing volcanoes in Hawaii. But Bird's first trip to America in 1854 ignited a lifelong lust for adventure, which continued well into her seventies.

Her letters home to her invalid sister, Henrietta, formed the basis of ten travel books. She wrote of riding alone through a bliz-zard with her eyes frozen shut, coming across a grizzly bear while trekking through the Sierra Nevada, and falling in love with a one-eyed Wild West desperado named Rocky Mountain Jim (who sounds even grizzlier). In Japan and China, she lived with and photographed remote tribes. While riding in northern India, she had a horse-riding accident and broke two of her ribs. By 1892, Isabella became the first female Fellow of the Royal Geographical Society, and her speaking tours would draw crowds of up to 2,000 people. Even well into her sixties, she was riding with Berbers in the Atlas mountains before being forced to her bed back in Edinburgh. When she died in 1904, her bags were packed for another trip to China. For Isabella Bird, travel was both medicine and liberation. "Travel is a privilege to do the most improper things with perfect impropriety," she said, which is probably what really cured her.

The dour-looking Isabella was not the only corset-wearing Victorian lady to transport herself to incongruously exciting surroundings. I love the photographs of Mary Kingsley, cruising determinedly down West African swamps in 1893, straight-backed and stiff in high-neck black lace and a bonnet while half-naked oarsmen from the Fan tribe row her along. "The cannibalism of the Fans, although a prevalent habit, is no danger I think to white people," she wrote. "Except as regards the bother it gives one in preventing one's black companions from being eaten." That cumbersome outfit came in handy occasionally. When she fell into a spiked game pit—dug by hunters to catch animals—she was saved by her petticoats. "Times like that you realize the blessings of a good thick skirt!" she quipped.

Mary Kingsley was remarkably undaunted by what was then known as "the Dark Continent" and Rudyard Kipling said of her, "She must of been afraid of something, but one never found out what it was." When a crocodile attempted to board her canoe, "a clip on the snout with a paddle" did the trick. A leopard that had ventured into her camp was discouraged by random items thrown in its direction. Which is probably how I would attempt to handle that problem.

Lady Hester Stanhope was just as brave when she ventured into the deserts of the Middle East after leaving Britain in 1810. At a public execution in Constantinople, Lady Hester was unfazed, only remarking that it was a shame the severed head was passed around "like a pineapple." She took a lover twelve years her junior—news of which scandalized society back home. Like Isabella Bird, Hester's doctor was the one who suggested a spot of traveling and even came along with her and later wrote her biography.

I also love the pictures of Gertrude Bell who, whether having a picnic surrounded by colonial soldiers, riding camels with Winston Churchill and T. E. Lawrence (aka Lawrence of Arabia), or exploring ruins at dusk in a fur coat, is always the impeccably dressed lone woman. And no wonder—her traveling caravan was said to include a full Wedgwood dinner service, a formal dress for evening attire, and a portable tin bath tub.

Criminally under-recognized today for her work in the Middle East, Gertrude's many achievements include negotiating with Winston Churchill and various Arab leaders to help draw up the then boundaries of Iraq, conducting hugely important archaeological digs in Syria and founding the National Museum of Iraq. She was also one of the first women to get a first-class degree in history in 1886, a decade before those Cambridge students were hanging effigies of women on bikes. But because of her sex she had to remain silent in lectures and was unable to graduate.

Inspired by the women I was reading about, I was still saying yes to nearly every travel opportunity that crossed my path. I'd accepted a commission to review a new five-star hotel in Bali, despite the fact that it meant eighteen hours of traveling and god knows how much carbon added to my footprint. I'd be staying there for a grand total of three days.

Previously, I would tend to avoid these kinds of trips as they're the junk food of travel journalism. I knew I should be saving myself for more nourishing assignments—the ones that inspired me and challenged me and where I got to see more of a place than just a hotel spa. But, just like junk food, these super-short long-hauls always seem like a good idea at the time. It's only later, when you're jet lagged and listless and alone in a hotel that could be anywhere, that you really start to question your life choices.

But because I was immersing myself in the stories of my fellow Victorian travel junkies, this somehow elevated even hotel reviews like this one to intrepid solo missions. While I wasn't quite packing my portable tin bath tub, as I lugged my suitcase on the train for the umpteenth time that month, I thought about Hester, who traveled to Palmyra in Syria in 1813 and took a caravan of twenty-two camels with her to carry all her bags.

I was also reading the bible for "adventuresses" as I embarked on this particular trip. *Hints to Lady Travelers* by Lillias Campbell Davidson was first published in 1889 to cater to a new market of women who were not only traveling but also reading about travel from the comfort of their chaises lounges. Aimed at women whose lives had hitherto been "unnaturally cramped and contracted within doors," it contains many sage words of wisdom, such as always carrying a small flask of brandy and not eating railway ham sandwiches. God knows what Lillias would've made of plane food. When I found my window seat and discovered I was sat in front of a screaming baby, I read Lilias's point that "a carriage specially devoted to babies and their guardians would be no bad thing."

As I pored over these tips on the plane, I thought about all the ways in which travel for me is less restrictive than it was for my forebears. *Hints* contains whole paragraphs on the best type of petticoat to pack and advice on provisions for horses on a driving tour. During a section on mountain climbing, Davidson tells us: "Let the skirts be as short as possible—to clear the ankles," but any shorter is "hardly consistent with the high ideal of womanhood."

But while modern-day travel for women solo tourers may have changed, some things remain starkly similar. Many of Lillias's hints are unexpectedly relevant for today's traveling woman—packing *is* "a ghastly preparation for a journey." I couldn't agree more. When I read her suggestion that "in traveling it is as well to take with one one's own tea," I checked my rucksack to make sure my stash of Earl Gray was topped up from my last trip.

When I finally landed in Bali, I tried not to think about having traveled 8,000 miles to review a single hotel or the fact that I'd imagined a tropical paradise, when it seemed to mainly consist of intense traffic and concrete shopping malls. Even Ubud, tucked inland and away from the more touristy beach areas, where Elizabeth Gilbert had her "love" epiphany in *Eat, Pray, Love*, looked like it had more Starbucks than spiritual awakenings.

But, let's face it, it's difficult to be churlish when you're lying by a pool drinking cocktails and eating club sandwiches "for work." From my sun lounger I lost myself in Lilias's suggestions for lady travelers. Although I can't say I was keen on trying her remedy for sunburn—"sour milk applied thickly at night"—and after swimming in the sea, I wasn't going to follow this particular tip: "It is a good plan to have the hair well washed with the yolk of an egg. Avoid hot water, which will cook the egg, with most unpleasing results."

On my last day in Bali, the hotel concierge, Danti, wanted to take me and the other journalists on a temple tour. Normally I hate religious tourism. It has always felt weird for me, a lifelong atheist, to intrude on other people's personal places of worship as if they're a tourist attraction. But in Bali it's impossible to ignore religion, even if you wanted to.

A tiny Hindu island amid the most populous Muslim nation on Earth, religion is everywhere in Bali. You see statues of the gods, shrines, and *canang sari*—colorful smoking flowers in foil trays—at every corner. Being Balian means a constant cycle of offerings and rituals. I watched people jump off motorbikes to make offerings right on the roadside. Prayer feels public and open and part of daily life here, so I agreed to come along.

I was expecting us to drive up to a stone structure, maybe a ruined one with a few vines and monkeys hanging about, like the temples we'd passed previously. But the hotel car pulled up in a car park on top of a cliff. We all dutifully followed Danti down a steep path to the beach, dodging the plastic bottles and straws which litter the coastline here.

First we heard chanting over the crashing waves and then we arrived at the temple. Just a few yellow and black and white gingham parasols stuck into the rocks. Offerings of fruit and flowers and smoking incense sticks had been placed underneath the umbrellas and monkeys were darting about helping themselves. Two men dressed all in white were blessing women in traditional Balinese dress—a white blouse, a yellow sash and a red and black patterned skirt. Danti passed us all some similar fabric to wrap around ourselves. The incense and the murmurings and the waves enveloped me.

I thought of Gertrude Bell's description of Damascus: "The air was sweet with the smell of figs and vines and chestnuts, the pomegranates were in the most flaming blossom." And her almost spiritual experiences on the water in Baghdad: "Have I ever told you what the river is like on a hot summer night? At dusk the mist hangs in long white bands over the water; the twilight fades and the lights of the town shine out on either bank, with the river, dark and smooth and full of mysterious reflections, like a road of triumph through the midst."

Before I really knew what was happening, I'd joined the line to receive a blessing. One of the priests cracked open a fresh coconut and poured the water all over my head while chanting some words that I couldn't understand nor hear over the sea. Through mime I was instructed to cup my hands and receive a garland of flowers. "You have to make a wish now," Danti told me. I felt the way I do on your birthday when you blow out your candles. Obviously it wasn't real. But you didn't want to waste it just in case.

The word popped into my head before I'd even had a chance to think about it: "A baby." I shocked myself. Where had that come from? I had always been fairly convinced that I didn't want kids. Choosing to be child-free, especially for women, can feel like an identity in itself, like you're in a secret club, and it was one I was happy to stay in. I loved my freedom and had never seen myself as particularly nurturing. But whether it was some kind of weird hormonal urge or simply because it now felt like I was a million miles from motherhood, a part of me was telling me that I wanted a baby. Which was a bit of a curveball, to put it mildly.

When we got back into the car, Danti effortlessly wove a red, white, and black thread around my wrist, called the *tridatu*. She told me the red thread was to symbolize creativity, the black to symbolize power, and the white was goodness. "We wear this to remind ourselves that life is not just one color, it's many colors," she explained.

Back in my hotel room, packing to go home, I felt disconcerted by what I'd wished for. This longing to be someone's mother felt strange and scary and with a life of its own, what the journalist Ariel Levy describes as "black magic." I thought about how telling it is that none of the British Victorian women I'd been reading about were mothers and only one had a husband—Isabella Bird, who got married aged fifty to her doctor, aka her travel enabler. Lillias Davidson, who went on to encourage even more women to see the world with her follow-up book, *Handbook for Lady Cyclists*, never married, preferring to live alone or with other women. We haven't come that far from the nineteenth century, when solo travel functioned as an alternative to familial and domestic responsibilities. It still feels to me that outside the roles of wife and mother, and after a certain age, women cease to exist in the eyes of society.

Certainly it was difficult for these Victorian women of means, having rejected domesticity and seen the world, to go back to a life of feminine conformity. Interestingly, Mary Kingsley refused to be aligned to the suffragette movement and resisted being called a New Woman. Gertrude Bell was a founding member of the northern branch of the Anti-Suffrage League, believing that the uneducated should not be involved in politics, no matter their sex. She thought that votes for women were against the natural order and held the view, rooted in her own personal success, that women were capable of achieving as much as men—even without the vote. Perhaps these female explorers had worked so hard to escape the domesticated fate of their "sisters" back home that, having done so, they had no wish to be allied with them.

Instead, Hester became the epitome of the crazy cat lady, living as a recluse in abandoned monasteries in what is now Lebanon with more than thirty cats. "Such dust!, such confusion!, such cobwebs!" wrote her doctor, Charles Meryon, when he visited her in 1837. She died alone and destitute in 1839 at sixty-three. Gertrude Bell also died alone and abroad. She supposedly never recovered from her married lover being killed in the Battle of Gallipoli in 1915 and, aged fifty-seven, took an overdose of sleeping pills in Baghdad.

On returning to London, Mary Kingsley couldn't forget the freedom she'd enjoyed in Africa, which "took all the colour out of other kinds of living." To a friend, she wrote of the stresses of coming home from a trip: "The majority of people I shrink from, I don't like them, I don't understand them and they most distinctly don't understand me… I cannot be a bush-man and a drawing-roomer. Would to Allah I was in West Africa now, with a climate that suited me and a people who understood me, and who I could understand." In the privacy of her Kensington home, she decorated her rooms with souvenirs from her journeys and wore African bangles. She was so desperate to get back to Africa, she enlisted as a nurse in the Boer War in 1900. At sea, she developed typhoid fever and died at thirty-seven.

I was interviewing a photographer on the phone at 1:00 a.m. from my hotel room in Bali (this happens a lot when you're freelance) when I absentmindedly clicked on my emails. Among all the random spam was an email…from Guy.

I looked at his name in the bold unread type for a long time. The shape of the letters. The effect they had on me. The photographer was talking at length about the light. My heart sped up. I couldn't get off the phone quick enough. The subject line read, "I know it hasn't been a few months but…" I took a deep breath and clicked on it. Guy's email—which I read and then re-read about 350 times—said that he just wanted to ask how I was and to tell me that he was still thinking about me. I didn't reply, even though a big part of me really, really wanted to.

CHAPTER EIGHT
Footloose and Fancy-Free

Teenage travels with Juanita Harrison

When I got back from Bali, I started to unpick this unexpected bolt of broodiness and what was really behind it. I'd always thought that motherhood wasn't for me. My favorite things in life are sleeping, spending all afternoon reading newspapers, and traveling by myself—hardly activities compatible with parenthood. I'd spent my life fighting against the feeling of being trapped, and having a baby was the ultimate trap, wasn't it? Deep down I think I'd always thought I was too selfish and broken to be "mother material."

Yet the societal pressure on a woman to have children is huge. It's in government advice to "hardworking families." It's in every advert you see on TV. It's in the first few questions a new person asks you, not realizing that "Do you have kids?" can often be one of the most emotionally charged and complicated things you can ask someone. I think for women, becoming a mother is still viewed as the default option and there must be some reason why you haven't—infertility, not finding "the one," or devoting your life to some higher purpose. If it's none of those, then it takes courage to go against the tide. I'd thought I was happy with my choice to stay child-free. So what had changed?

For some time now I had felt caught between two stages of life. I seemed too old to go out dancing until it got light as I had as a teenager, and I no longer wanted to throw myself completely into my work as I'd done in my twenties. So what came next? I remember being in a park in Lisbon, sitting alone with my book. I found myself between a group of twentysomethings smoking weed and listening to music and a group of families having a picnic. It seemed to perfectly sum up this feeling I had of being in limbo land.

No doubt tied up in this desire to become a mother was my relationship (or lack of it) with my own mother. She had left when I was fifteen to live with her new husband, and although we had tried to maintain some semblance of a bond, after many years of fractious and difficult meetings, I'd decided I no longer wanted her in my life. Having not spoken to her in months—part of my new post-divorce plan of not doing what other people expected of me—I felt the first inklings of what kind of mother I might want to be. It was almost as if I had to sever all ties with this dying relationship to allow the space for my own fresh shoots of motherhood to grow.

Coincidentally, around this time I happened to stumble on a quote from Freud, who has a lot to say about coincidences. He thought that the psychical foundation of all travel was the first separation and the various other departures from one's mother, including the final journey into death. Which explained a lot.

I had not replied to Guy's email. I knew that if he and I started talking again, we would be straight back to the heady teenage lust days of before. It was only now I was over my heartbreak that I realized how in a bubble I'd been when we were seeing each other. It all felt wild and irresponsible and intense—and also very unsustainable. We once got kicked out of a hotel for staying in our room three hours after checkout, having been up all night drinking champagne in the bath. When we were apart,

I'd often spend hours glued to my phone, re-reading messages from him, gazing at pictures of himself he'd sent me. We once had a thirty-minute conversation just using emojis.

Whenever we were together, well, we spent all our time in bed. Until eventually one of us got thirsty and someone had to get up and make drinks. I think I subsisted mainly off cups of tea and martinis during that time. Which was fun and all, but every other part of my life suffered as a result. Work deadlines were missed; friends didn't get phoned back. At the time I'd been able to blame my flakiness on my divorce, but I was in a different place now. Besides, the way things were with Guy didn't feel like any sort of basis for a healthy long-term relationship, let alone a family—if that was what I wanted. I had finally started to look back on our affair with a fondness rather than a crushing physical pain. Until Guy, I'd forgotten how powerful being in that first rush of love can feel. It really is just like being fifteen again.

Our teenage years are when some of us get our first taste of solo travel. This could be a gap year—whether that means digging wells in Uganda or swigging beers in Australia. For others, it's an exchange trip, where you're sent to live with a host family in another country. As you'd expect from sending teenagers to live with strangers in a totally different culture, things don't always go to plan. The best story I heard was of a Moroccan family slaughtering a goat right in front of their fourteen-year-old guest to welcome her. It was a pity she was vegetarian. Just like your first boyfriend, you never forget your first travel experience—even if, like the boyfriend, you'd never want to repeat it.

The trips you take with friends when you're a teenager can also be very formative. Aged seventeen, I went on an ill-advised eighteen-to-thirties-style holiday. There were fifteen of us, all girls. Still blinking from months of studying for our exams, all wearing white T-shirts with our nicknames emblazoned on the back in pink letters. "McMuff," "Little Jo," um…"Willsy." But we didn't need great nicknames. We had just discovered that you could get served free Baileys on an airplane.

This was our first real holiday without parents, teachers, or Brownie leaders. We had been unleashed. Descending en masse to Malia, a "party hotspot" in Crete, we were confronted with a Disneyland of debauchery. I can recall scenes that wouldn't have been out of place in the sleaziest hip-hop video or a Hieronymus Bosch painting. We soon realized that we were a long, long way from our all-girls' grammar school in suburbia.

As anyone who has been cooped up at a single-sex school will understand, the mere sight of a potential mate, no matter how unattractive, sunburned, or drunk, is enough to send you into a frenzy. Which is probably why one of our number (a usually shy, bookish girl) spent the first night stripping down to a thong and dry-humping a holiday rep on stage, while the crowd screamed and cheered, held up score cards to give her a mark out of ten, and then hosed her down with giant super-soakers. Something about the combination of sun, sea, and spirits so strong they'd be illegal back home made a group of normally sensible, swotty students turn into girls gone wild.

We endured seven long days of fishbowls, foam parties, and heavy petting in swimming pools. I came home with a mouthful of ulcers (which I hope was due to general rundown-ness rather than an STD) and possibly also scurvy, because the only fruit or vegetables I consumed all week were on the rim of a cocktail glass. There were some sobering moments. Four of our number were chased down a backstreet by a cab driver screaming, "You are the English! You are here for the sex! I want the sex with you!" Luckily they managed to run back—teetering in sky-high wedges—to our decrepit hotel unscathed. I woke up more than a few times with absolutely no knowledge of how I'd got home. Or why there was a plastic shopping bag of vomit next to my bed.

I learned some other important truths on that holiday. Not least that drinking unknown quantities of a substance pertaining to be "Sex on the Beach" all day, in the sun, will make you projectile vomit for forty-eight hours straight. Like many teenage trips abroad, it was a first, intoxicating sip of the big wide world.

It was also an early lesson in not necessarily having to do what every-one else is doing. For me, going on a "girls on tour" trip at seventeen felt like a mandatory rite of passage I had to go through. But, like bachelor-ette parties, freshers' week, and Christmas, the things that should be fun are often anything but. I wish it hadn't taken me twenty more years to figure that one out.

Although I never had a gap year—I was itching to get out of my dad's flat and get to university—I did go on a monthlong trip to Thailand with my three best friends from school in the summer holidays after my first year of uni. Despite the Malia experience, and a few terms of tertiary education under our belts, we were still very naive. So naive in fact, that when we got to the airport and saw a Starbucks, we were so excited that we promptly sat in it for several hours, despite the fact we hadn't even gone through security yet, and thus only narrowly avoided missing our flight. We had about fifty-six As between us, but working out airports wasn't yet in our skill set.

My diary from that trip reveals just how little I knew about the world but that I fancied myself as a travel writer even then, with clichéd prose about the Khao San Road being "a blur of neon signs…fumes and heat and incense overpowering." We all lugged around backpacks that weighed more than we did, wore fisherman trousers, and played pirated Jack Johnson CDs off our Sony Discmans. We went to full moon parties and drank buckets of Thai whisky and Red Bull. Every hostel had damp mattresses, cracked walls, and cockroaches, which we dispatched with a thud of our weighty Lonely Planets. If we did go to any religious, cul-tural, or historical sites, my diary has no mention of them. Although one entry records these culturally sensitive and spiritually inquisitive words: "Still not serving alcohol due to that Buddhist religious thing."

One encounter we had in Thailand looms large in my memory. We were in a bar called "Rolling Stoned" in Bangkok, where a Thai version of Mick and co. played covers, when a hippie-looking American woman started talking to us. Back then she seemed impossibly old, but thinking about it now, she probably wasn't more than forty-five. "Girls," she said, looking us dead in the eye and adopting the solemnity of someone delivering an eternal truth. I can still hear her voice now, nearly twenty years later. It sounded ravaged by experience and probably also Thai cigarettes. "There's six million miles of cock out there. You've got to go out there and get your share." You can imagine how many times we repeated that catchphrase to one another (and occasionally still do).

My first travel experiences were a booze-fueled, carefree whirl, but I think times have changed. Many of today's teenagers taking their first trip abroad seem less driven by hedonism and more by Instagrammable moments. Or perhaps they're a more discerning generation, looking for immersive cultural experiences rather than dodgy cocktails.

I'd been reminiscing about my own teenage travels lately because I'd just discovered Juanita Harrison. I was now reading any travel memoir by a woman that I could find, and when I heard about *My Great, Wide, Beautiful World*, published in 1936, it immediately took me back to my own first tentative steps.

Juanita Harrison was just sixteen when she set off on her adventures in 1906. "I will sail far away to a strange place," she wrote in the preface to *My Great, Wide, Beautiful, World*, one of the earliest examples of autobiographical travel writing by an African American woman. "Around me no one has the life I want. No one is there for me to copy, not even the rich ladies I work for. I have to cut my life out for myself and it won't be like anyone else."

Juanita was born in Mississippi in 1890 or 1891, where opportunities for Black women were scarce. This was a time of segregation, mass lynchings, and race riots. Juanita had only had "a couple of months" of schooling before she left education at ten years old and began working as a maid to support her family. Yet she resolved early on to swap a life of domestic drudgery for seeing the world. Juanita went on to work her way around twenty-two countries alone. She learned French and Spanish. She visited the Moulin Rouge, the Taj Mahal, the Dead Sea, casinos, spas, and circuses. She was involved in a train wreck where she cradled a dying woman and was assaulted numerous times, both sexually and physically.

We'd never know about any of this had it not been for one of Juanita's employers suggesting that she should put her travel experiences into a book and then helping her to get it published. Undaunted by her lack of literacy, Juanita vowed to make this book "just as I have written them misteakes and all. I said that if the mistekes are left out there'll be only blank." Which is a rather lovely metaphor for life, too.

My Great, Wide, Beautiful World is a masterclass on how to travel solo, at any age. The teenage Juanita positively relished being "the lone woman" and found that once people discovered she was on her own, everyone wanted to talk to her. "How well we can carry out our plannes if its just yourself," she noted. "When you find the Places alone you enjoy it better."

It's also full of useful tips if you're on a budget. From the first page, she's pilfering soaps and towels from her cruise ship and sneaking along with tour groups in Rome. She preferred to stay with families rather than in hotels, was a big fan of "loafing" about doing nothing (often the best way to soak up a feel for a place when you have no cash) and offers this truism: "The First and second class are the same the World over but it is the 3rd class that are so interesting."

Because Juanita had the luxury of time but not money she would happily "go hungry a few times to Visit an extra Town" or wait for a cheaper train to see "more interesting people." She was also more than willing to accept a discount based on the fact she looks like a "poor lonely woman…but there never were one less lonely than I am."

Despite being short on funds, she didn't often go hungry because she was, in her own words, "an expert marketeer," heading straight for the food market when she arrived in a new place. "One of the thing I enjoy as I go to each country is how the breakfast of the working people change," she wrote. "Italy it is little crisp fried dishes. Bucharest Rumania little suckling pigs meat. Cairo red beans young green onions boiled eggs and oilive oil Seville fried rings of bread broned rich and hot, stewed trip with tomatoes parsley and wine five or 4 cents."

Her mouthwatering descriptions of "crisp fresh sardins and oranges and wine" in Seville prove that often it's not Michelin-starred menus but the simple tastes and pleasures that are the most evocative of a place. "Of all the things in the world Ice cream can only give me an elegant feeling," she wrote. In Spain, she also got a taste for bull fighting and declared that "Bull fighting and ice cream are the two best things on earth."

I love that Juanita was proud of how little money she spent on her trip, as some of my best travel experiences have been my most frugal. After I'd finished my year abroad in California, I went to live in New York. I had about one hundred dollars to my name and no job, yet moving to one of the most expensive cities in the world didn't worry me for some reason. Perhaps I'd watched too many episodes of *Sex and the City* and *Friends* and thought you could live a lovely lifestyle by occasionally writing a column or serving a few cappuccinos.

I subletted a tiny basement room in an area of Brooklyn that is now full of man buns and flat whites, but back then was very much clouds of weed smoke and unsavory-looking people hanging out on street corners. There were bars on the window of my room, no curtains, and no air conditioning, which in August in New York felt like one long Bikram yoga class. Because it was so hot in my room, I had to sleep naked with nothing covering me. One morning, I woke up to find an old man taking photos of me through the bars of my window. I was so tired I just pulled a sheet over me and hoped he'd go away.

I took a series of random and terrible jobs that were elevated by their New York novelty into being quite fun. I gave out fliers in Madison Square Garden. I wiped down yoga mats at a studio in the East Village. I poured champagne at fancy parties in the Hamptons. But I spent most days just wandering about the city and found out that New York is actually a really good place to be aimless and broke.

I came up with a brilliant hack where I'd go into coffee shops and order a caramel latte and then when they made me an iced one I'd say, "Oh sorry, I meant hot," and vice versa. I worked out that they would normally end up giving me both and I'd then drink them in quick succession, sometimes feigning a theatrical eye-roll of annoyance that they'd got it wrong.

I remember stumbling across a performance of *Macbeth* on a street corner, sneaking into screenings of old films in Bryant Park, subsisting solely off pizza slices that were bigger than my head and going to dive bars where beers were a dollar. I pretended I didn't know about the tipping the bartender for every drink thing because I was British.

I witnessed so many "New York moments"—probably because I was just wandering around the hot asphalt for hours, high as a kite on sugary caffeinated drinks. I saw kids splashing about in fire hydrants in my neighborhood. I played chess with old men in Central Park. I sauntered into free museums and stood in front of important art and waited for it to make me feel something. Sometimes it did, but that might have been the coffees.

One night I did treat myself by going to a fancy Italian restaurant because Sam had come out to visit me. We ordered the cheapest pastas on the menu and a bottle of a wine that was considerably less expensive than all the others. "This is a dessert wine, ma'am," the waiter told us, looking embarrassed on our behalf. "That's fine," I answered breezily. "We want that one." Of course, we drank every single saccharine drop of it. And then I promptly vomited up my spaghetti carbonara all over my boiling hot basement room when I got home. I remember looking down at the pile of vomit and thinking that this was fifty dollars not well spent.

Throughout her travels, Juanita displayed a similarly undimmable enthusiasm and positivity that only someone experiencing the world for the first time can muster. She was undaunted by the weather—"I just love to slop around in the rain," she said in London—or by having to sleep in a bathroom for four nights in Nice while she waited for a room to become available: "I enjoyed it the tub are large and plenty of hot water I had 2 and 3 baths every day." She traveled light, "one suite case is one to many," giving possessions away as gifts to people she met and buying clothes more suited to the local environment as she went along.

Juanita's observations are prosaic but often surprisingly profound—full of youthful, wide-eyed naivety and impatience. "It's nice to be in gay Naples after Churchic Rome," she wrote. In Spain, she wished "for another pair of eyes my two were so busy." In Bethlehem the Old City is described as "a real cross word Puzzle," and she observed, "There will never be any content here." Almost one hundred years of religious strife later, and that still seems to be the case.

In Sri Lanka, she took an accepting view of religion: "The Hindoos had the same right to think they were right as the Buddish and as we—how did we know any one were right." The tradition for tattooing women's faces in India inspired her to note that being a woman often means enduring pain for aesthetics: "Where they suffer a few days from that, we suffer for life from corns made by Tight shoes."

Travel allowed Juanita Harrison to transcend race and class and experience the world in the way that most women of her age, in her time, could not. She defied attempts to categorize her and was variously mistaken for Italian, "Aribian," Jewish, and Chinese. "I am willing to be what ever I can get the best treatments at being," she remarked pragmatically. She was a master of meeting locals and getting stuck in with whatever they happen to be doing. Of joining in the dancing at a fair in Czechoslovakia, she wrote, "Of course I was not a tourist. I was one of them."

She was also something of a man magnet—"I am auful foxie," she commented—and adept at giving admirers the slip when she didn't want to go home with them, sometimes landing an "upper cut under his chin" when boys got "fresh." Her delight in handsome men around the world and her "snapy eye for flirting" is often unintentionally hilarious. While on the shores of Varanasi in India she noted, "A good Flirtation was just the thing after the Funeral Pyres."

It's only when she arrived in Hawaii at the end of her travels that Juanita recorded her first experience of any racism. While looking for a job she was told that "white People here want only Japanese help." But she wasn't discouraged: "As if any nation can keep me from getting a job." She eventually settled on the island of Waikiki, in her own tent, in "a privat Place yet free good and cheap." Juanita spent her days swimming, eating pineapples, and recalling her adventures, letting her "mind travelle to Zeurich Switzerlan Zufolo Iseral Philisitine Sweden and Madras, Ceylon and Smiling faces stands out before me." The last page of her book reads: "I'll get a serfe boad and Take a few Hula lessons." Which sounds pretty good to me.

It's easy to romanticize travels past, especially teenage ones, in the way we wistfully recall old lovers. But I do think there's a lot to be said for giving yourself budget constrictions while traveling, especially when you are solo. Luxury travel is boring and sanitized and can be quite lonely. The Four Seasons in São Paulo is the same as the one in St. Petersburg. When in a new place, I've had much more fun walking around the streets on my own, going where I pleased and meeting people, than I have being ferried around in a slick air-conditioned private car with my own guide.

No record of Juanita's life exists after the publication of her book. She appears not to have written any more in her lifetime (probably too busy hula-hooping) and details of her death are unknown. Yet there's one picture of her that I found online. It featured in the *New York Times* review of her book on May 17, 1936 (the reviewer's verdict of *My Great, Wide, Beautiful World*? "Its humor is so rich, its comment so racy."). The photograph is black and white and shows Juanita in Hawaii with her toes in the sand, a wave gently breaking. She's wearing a checked sarong and a slouchy vest top, and she's throwing her head back laughing.

Looking at her photo, reading her book, I think Juanita is in many ways the ideal solo traveler. She's not so po-faced and obsessed with having "authentic" experiences that she doesn't have fun. But she works hard to immerse herself in a place, to talk to the local people, and to experience it fully. I wish my teenage travels—and indeed my travels now—could be as carefree and effortless as hers. She is also completely right when she notes: "I have reversed the saying of Troubles are like Babies the more you nurse them the bigger They grow so I have nursed the joys." After reading *My Great, Wide, Beautiful World*, I vowed not to focus on all the things that I'd done wrong over the last few months, or all the things that I'd lost, and instead to nurse the joys.

I was also still nursing my bracelet from Bali, even though I felt a bit like a "Gap Yah" student who refuses to take off their traveling jewelry or *that* music fan who keeps their festival wristbands on for way too long. But I wanted a reminder of the wish I'd made, standing by the Indian Ocean with coconut water dripping down my face. I did want a family, I knew that now, and looking at this bracelet and reading Juanita's book gave me faith that things would all work out if I gave them some time.

HOW TO TRAVEL SOLO ON A BUDGET

There's a myth that being alone can make a trip more expensive, but thankfully the days of hotels and tours charging a single supplement are, for the most part, over. That said, if you're not sharing the cost of hotel rooms/taxis/car rental with someone, traveling by yourself can add up. Here are some ways to make your money go further.

1. **Rediscover the hostel 2.0.**
Forget thoughts of smelly dorm rooms for teenagers, moldy communal showers, and threadbare bunk beds, there's now a whole new breed of hostels that come with meditation rooms, yoga studios, and hot tubs. You can always opt for a private room with a bathroom (if the risk of bunking up with six Australian teenagers on their first holiday is too much to bear), but many dorms are women only and can be wonderful. There is something comforting about the communal solitude of strangers. Most hostels have kitchens so you can cook and make your own food—another big way of saving money. If you're not into hostels, then homestays or a room in an Airbnb are also great options when you're alone. Not only are they much

cheaper than hotels but they also make you feel like more of a local and less of a tourist.

2.   Find the free stuff.

With a bit of research, you can find fun things going on in any city that you don't have to part with any cash for. Many museums, galleries, and tours have free entry or "pay what you can" days. The mecca of free things has to be Washington, DC, which has more than seventy museums, most of them costing absolutely nothing to enter. If you can't find anything free where you are, one of my favorite things to do in any city is to people-watch in a park.

3.   Look for discounts.

Because you can often be more flexible when you're traveling by yourself, it's possible to get bargain flight deals. Set up a price alert on booking websites such as SkyScanner or FlightNetwork so that you get an email when the flight price drops, or book something really short notice for a fraction of the price. Call up on the day and ask if hotels can offer a last-minute rate. Remember that hotels lose money when rooms are empty, so they would rather have an occupied room at a discount than an empty one.

4.   And sometimes ask for other discounts, too.

You need a lot of chutzpah for this, but I have a friend who will simply get chatting to staff in a shop and then ask "Are there any discounts today?" or "What's the best deal you can do on this?" She swears that she typically gets 10–15 percent off. In some places—street markets, for example—haggling is not only appropriate but expected. The general rule of thumb is that you should half their first offer and work up from there. Have in mind a price you're willing to pay and stick to it. Also bear in mind (or have an app on your phone to tell you) the exchange rate. It's easy to get confused and caught up in the excitement/ stress of haggling and end up paying more than the cost of your whole trip for a rug that you didn't like that much anyway.

5. Work it.

If you're short on money (and even if you're not), consider incorporating some work or volunteering into your trip. There are many places that provide room and board in return for a certain amount of unpaid work, such as WOOF-ing (working on organic farms). Be prepared to actually work, though. You have to be on board with mucking out animals or sticking your hand in a beehive and won't be able to just sunbathe all day long. But not only will you learn new skills, it can be a rewarding and inspiring way to experience a place (and get a cheap holiday). The website Workaway connects travelers with volunteering and work opportunities around the world.

6. Take a break from the norm.

Also known as *undertourism*, by thinking outside the box on your destination, you can save stacks of cash and skip the crowds. The typical tourist spots tend to be expensive, so by choosing some-where a bit off-radar, everything (accommodation, food, trans-port) will instantly be more affordable. For example, instead of Amsterdam, consider Rotterdam. It has a cutting-edge art and electronic music scene and a cool film festival held in beautiful old theatres. New York can be a cash-hoovering monster, but Philadelphia has stylish speakeasys and world-class art muse-ums, too. Think Albania, not Croatia; Puglia, not Tuscany. Also, consider destinations that have recently been affected by natural disasters or terrorist attacks (provided government advice is that it's safe to go there). By supporting economies that rely on tourism, your presence will be so much more appreciated—and your money will stretch a lot further, too.

7.   Eat where locals eat.

As a general rule, I find that the cheaper the eat, the more fun the experience, especially when you're solo. Don't discount the joy of going to a market or supermarket and just making a picnic (an indoor one if you're not somewhere warm). Try to avoid tourist traps where prices are hiked and the food is often average. Places with menus in multiple languages right by a popular attraction are usually to be avoided. Everyone knows an amazing and affordable place to get food in their hometown and they usually love telling people about it. Ask your taxi driver, your tour guide, the random person you chat to on the bus where their go-to cheap eat is. Mine is a salt beef bagel from the Beigel shop on Brick Lane, if you're interested—£3.90 for hot juicy beef, thick slices of pickle, and a slather of mustard on still-warm, chewy bagel. *And* it's available 24/7. See you in the queue.

8.   Get public transport.

Don't be bullied by people into thinking you can't manage it. If you've navigated public transport in a major city before, chances are you can do it wherever you are. Traveling by taxi is great when you are knackered or have a lot of luggage, but if you can, it's much better and cheaper to travel like the locals do. This is also another reason it makes sense to pack light (see packing tips on page 106). If you are getting a taxi, always negotiate the fare up front (and in many places be prepared to haggle) as nothing says "hike up the meter" like a wide-eyed newbie lugging a giant suitcase. The ubiquity of Uber (for good or ill) has put paid to a lot of dodgy cab schemes, and most airport taxi ranks have a fixed price, but it's good practice to ask what you'll be paying before you get in and in general to have an idea of what things should cost so you at least know when you're being charged a "tourist tax."

9.   Take your own water bottle.

You'd be surprised at how many people's good habits at home go out the window when they're traveling. Not only does a refillable bottle save you a lot of money (trust me, those overpriced Evians add up), but all that single-use plastic ending up in landfill/on the beach/being burned isn't a great thought. In places where the tap water isn't drinkable, be really sure that the water you're fill-ing your bottle up with is 100-percent safe. Most hostels, hotels, restaurants, and cafes now have a water station for guests to refill their own bottles. In some countries, you might find that you have to buy a big ten-liter bottle of water at the supermarket, keep it in your room and refill your bottle from that every day, but you'll still save money/the planet.

10.   Think about staying somewhere longer.

The trips where I've spent the most money are the ones where I've moved around a lot. I'm not sure if this is because I'm more tired, so I tend to buckle when I see the appealing lights of a taxi winking at me, or if I always need to buy that latte or another brownie to keep going. If you stay somewhere longer, not only do you get to feel more like a local and get to know the cheap places to eat and things to do, you're also more likely to go to the supermarket and cook.

11.   Have a big breakfast.

The old saying that you should breakfast like a king, lunch like a prince, and dine like a pauper is particularly true when you're traveling alone. Breakfast is a brilliant meal to linger over when you're by yourself—you can plan the day, and it's often included in your hotel rate, so if you're on a budget it pays to fill up first thing in the morning. Lunch is better out and on the go, either from a market or look for restaurants that offer cheaper lunch deals. Then cook dinner if you have a kitchen or just have a small snack at a bar with a glass of wine. This also avoids the "argh, look at all the couples having romantic dinners" moment that can strike when you're by yourself.

12. Mix it up.

Whether you book one fancy dinner then eat instant noodles for the rest of the trip, or you decide to stay somewhere lavish for a single night and then couch surf your way around the rest of the country, switching up your spending makes you appreciate both aspects so much more. And remember, you don't have to be staying at the best hotel in town to enjoy its bar, spa, or restaurant. Just go as a day guest and lap up the ridiculous splendor and then make your way back to your budget hostel dorm room feeling pampered.

13. Be money savvy.

There are now many banks that allow you to withdraw cash internationally without any extra fees or charges. Try to avoid changing money at the airport if you can help it (it's the worst exchange rate and the most exorbitant commission). Most airports will have an ATM, so getting cash on arrival is nearly always your best and easiest option. If you're traveling outside Europe, I also find buying a cheap SIM card at the airport can save you a lot of money in roaming and data charges and avoid a nasty bill on your return.

14. And remember...

Even if you are penny-pinching, travel is one of the best things you can spend your money on. A study from Cornell University found that money spent on experiences brings more lasting happiness than possessions. Because we "adapt" to physical objects, the joy of owning them wears off. But the buzz you get from remembering a trip only increases with time. So if you feel bad about splurging on flights or dropping a load of cash on a hotel for a couple of nights—don't. It's scientifically better for you than saving up for that new iPhone/dress/house.

CHAPTER NINE
The Great Escape

Getting away from it all like Robyn Davidson

Traveling started to feel different now that my life was calmer. Although I was still doing plenty of solo trips, they didn't feel quite so frenetic. Because I was no longer so restless, I didn't pursue them pathologically like I had before, filling the hole by planning yet another escape. I felt happy in London and had even been on a few fun dates, but still I found I couldn't stop thinking about Guy. He would pop into my head when I least expected it. Or when I was having a weak moment I would look wistfully at his social media. I wondered what he was doing and whether he was still thinking about me, too.

Eventually I started to question why I had to be so strong all the time and fight my feelings. Maybe there could be a way back for us? I decided to send him an email—*What was the harm?!*—an impressively breezy-sounding email at that. He replied almost instantly and we started talking again. A little at first and then every day.

We arranged to meet up when we were next both in the country at the same time. Which in practice meant that when the day rolled around a month later, Guy had just got back from a surfing trip in the Seychelles and was looking tanned and so ridiculously handsome that it actually felt quite unfair. I instantly realized how much I had missed him. It felt strange to see someone I had imagined so many times in my head in the flesh. A bit like when you finally see that famous landmark you've scrutinized a picture of so many times in your guidebook.

We did a lot of catching up. He really did seem sorry for what had happened. I realized that I'd been holding back from telling him how I really felt about him because he didn't fit in to some arbitrary checklist I had in my mind about the type of person I should be with. I had discounted what we had as pure lust, when in fact I could see now that it was much more than that—it just hadn't looked the way I'd imagined.

As we started to get to know each other all over again, it was all the good parts from before but without any of the anxiety. We'd ditched the secrecy and the high drama of our illicit too-soon-after-the-end-of-my-marriage affair and replaced it with something that seemed like it might be even more exciting. We told each other we loved each other. I raised my concerns about the more hellraising aspects of his personality, and he confessed that he'd been playing them up because he thought I was into bad boys. He told me he'd never really known how I'd felt about him and thought he was just a rebound for me, and I conceded that I could see why that might've been the case because, to be fair, I had thought that, too.

Although I tried to force myself to take things slow, within weeks we were inseparable again. I was floating with happiness (I must've been truly sickening to be around). I couldn't remember ever allowing myself to feel this way before—I was too cynical, too guarded. But having been to the depths of sadness, I felt I was now entitled to a bit of the other extreme. Plus, it was summer, which meant long hazy afternoons lying in the park and eating ice cream. Simply being in London and in love felt like one long holiday. I realized I had no desire to go anywhere without him. This was the first time in a long time that I hadn't felt the craving to get away from it all.

Luckily, most of the trips I had arranged, Guy could come along with me. We had an amazing few days in Matera in southern Italy, a city which looks like it's been cut and pasted out of a storybook. I know they say that Paris is the city of love, but I reckon Matera's secret caves and windy cobbled streets put up a pretty good fight. We drove Guy's vintage Porsche down to Devon, where he educated me in the lesser-known works of Guns N' Roses and we swam in a freezing cold sea under ancient stone arches on the Jurassic Coast.

It felt weird getting used to traveling with someone again. I'd got so good at the lone ranger thing, and now there was this other person by my side. One who insisted on carrying my bags, no less. It was weird but also wonderful to have someone to share all these new travel experiences with. I didn't even mind that I couldn't starfish in the huge hotel bed or that he slowed me down when going through security at airports (I have my routine so finely tuned and streamlined by now that mere mortals can't keep up).

I felt the inherent rightness of things slotting into place. Being with Guy felt effortlessly fun and work was going really well—I'd been asked to write a weekly column for a national newspaper (finally, my Carrie Bradshaw dreams coming true!). I was doing a photoshoot for my column when I happened to mention all the female explorers I'd been reading about. The young Australian photographer's assistant asked me if I'd heard about Robyn Davidson. I hadn't but as soon as I got home I bought her book *Tracks*. As I started reading it, I realized that for some women, a solo adventure doesn't mean going away without friends or a partner; it means going somewhere without any other people *at all*. Traveling somewhere not just without company but devoid of any signs of humanity—houses, shops, trees, road signs, roads. What Martha Gellhorn called "falling off the map."

When Robyn Davidson was twenty-five, she left her home in Sydney for the remote town of Alice Springs. It was 1975 and she had a dog, six dollars, and a suitcase of inappropriate clothes. She describes being "vaguely bored with my life and its repetitions," a feeling I could entirely relate to at so many points in my life. *Tracks* begins by describing another state all too familiar to me: "I experienced that sinking feeling you get when you know you have conned yourself into doing something difficult and there's no going back." And this is all before she'd even started on her 1,700-mile journey across the desert.

Despite never having actually seen a camel before, Robyn had a "lunatic idea" to trek across the Australian outback with them, from Alice Springs in central Australia, all the way to the Indian Ocean off the western coast. She was searching for meaning and an escape from modern life. She had been reading a lot about Aborigines and saw traveling through the desert as a way of getting to know them. She had "never held a hammer, changed a tire, or used a screwdriver," but soon she was designing and building saddles and firing guns. Before her trip even began, she battled "the Australian cult of misogyny," working in a pub where men pissed up against the bar, told her she'd been nominated as the next town rape case, and left a lump of human excrement under her pillow. While learning how to train camels with a "maniac" who owned a camel farm, she was forced to work with no shoes and couldn't sleep at night for the pain in her "swollen, punctured, infected feet."

Many people called her trip a suicide mission or a publicity stunt, but its extremity was part of the appeal. Robyn wanted "to be alone, to test, to push, to unclog my brain of all its extraneous debris, not to be protected, to be stripped of all the social crutches." She relished having "no more loved ones to care about, no more ties, no more duties, no more people needing you to be one thing or another, no more conundrums, no more politics, just you and the desert, baby!" And who can say that doesn't appeal?

When she eventually set off into "rough uninhabited country…no one and nothing for countless miles," she endured poisonous snakes, injured camels, dehydration. She was forced to eat her dog's biscuits when supplies ran low. Walking mile after mile through endless dunes she entered "a new time, space, dimension. A thousand years fitted into a day and aeons into each step."

When you follow her route on a map, the sheer scale of it is truly staggering. And there's the fact that she did it all alone, with only four moody camels and her dog Diggity for company. With distances like these, and the sheer remoteness of the landscape, taking a wrong path or finding a water source has dried up is the difference between life and death. Over nine months she walks nearly 2,000 miles west to reach the Indian Ocean—the equivalent of trekking from London to Marrakech. Even just reading about it is truly terrifying. "Living on one's nerves and expecting every moment to produce a horrendous catastrophe is one thing—doing it in 130-degree heat is quite another."

National Geographic sent a photographer, Rick Smolan, to capture points of her journey on camera in return for much-needed sponsorship money. Robyn was deeply conflicted about allowing him to join her but, though at first he was an annoyance (from New York, Rick comically has no idea about the desert and brings with him an inflatable dinghy), she and Rick become lovers while in the desert and, eventually, lifelong friends.

Robyn's desire for the experience to change her was so strong that she records that she felt annoyed after two weeks that she was "exactly the same person that I was when I began." But after a while she realized how the environment was altering her. She noticed a shift, as the very boundaries of who she was began to "melt" as "the openness and emptiness which had at first threatened me were now a comfort."

I've noticed this shift on my own solo trips. While for the first few days without anyone to talk to, it can often feel scary to feel this space open up in my thoughts, soon I start to actively crave that liberating feeling I get from being immersed in a place. It can be unbelievably freeing, when everything else falls away, to be where no one knows you or expects anything of you or wants you to look or act or behave in a certain way.

Mr. Eddie, an old Aboriginal man whom Robyn walks with for a month, taught her to let go of constructs such as time and structure. She developed a new state of mind and started to shed her former self and all her social niceties. Soon she was walking naked, period blood running down her legs, having long kissed goodbye to "the disguises and prettiness…the horrible, false, debilitating attractiveness women hide behind… In my own eyes I was becoming sane, normal, healthy, yet to anyone else's I must have appeared if not certifiably mad then at least irretrievably weird, eccentric, sun-struck and bush-happy."

This part of *Tracks* struck such a chord with me. I couldn't help thinking back to the early stages of seeing Guy—and, in fact, the initial stage of all of my relationships—and the veneer of falseness I cultivated to make myself seem desirable. I would go to great lengths to seem pretty and perfect and—crucially—give the impression that this prettiness and perfection had been obtained without any effort. I would always be wearing my sexiest—and often most uncomfortable—underwear whenever Guy came over. My body would be waxed to Barbie-doll levels of smoothness. I wanted to be "the Hot Girl" at all times and it was exhausting.

During my teenage years, I remember waking up before a boyfriend to apply my makeup and then getting back into bed so that he would think I "just woke up like that." I sometimes wish I could have all the hours back that I spent fake-tanning and hair-straightening during university. Although I'd moved on a lot since then, I was still making a hell of a lot of effort to disguise the real me, to appear well groomed at all times. So perhaps this had been part of the pull of solo travel for me—that I could be messy and sweaty and mosquito-bitten away from anyone who knew me who might judge me or care.

We're so used to seeing women plucked and starved and painted and spray-tanned and hair-dyed that to "let ourselves go" (Go where exactly? Well, quite a lot of places in my case.) and allow ourselves to remain au naturel can feel liberating. When you internalize these inherited and unrealistic ideals of female beauty, you can find you have become that woman in Iceland who didn't swim in a lagoon because she had blow-dried her hair. I never want to be that woman. Thankfully, this time around I felt comfortable enough with Guy that I could drop the facade. Although I hadn't quite reached Robyn Davidson levels of wild woman-hood, I'd found my own version of "bush-happy" (so to speak). I let my guard down and my body hair grow. I stopped wearing makeup and pre-planning outfits. Guy hadn't run a mile.

The press picked up on the story of "the camel lady," and Robyn Davidson became a feminist icon. But when the reporters tracked her down and approached her in the desert, it had been so long since she'd seen people that they looked like "invading war lords." When the trip was over, she went to New York with Rick and was overwhelmed by the "cars and the telegraph poles and questions and champagne and rich food" to the point of actual vomiting.

Robyn realized some fundamental lessons while crossing the desert alone—words that could sound glib but actually feel heartfelt after the journey the reader has been on with her in *Tracks*. She learned that you can't have true freedom without risk, that "you are as powerful and strong as you allow yourself to be, and that the most difficult part of any endeavor is taking the first step, making the first decision." Which, although I've never trekked across half of Australia with camels, rang true for me too. The hardest part in all I'd been through was realizing I could no longer stay with Sam. That one decision had set off so many huge and frightening changes in my life, but that first step had definitely been the scariest.

A different kind of desert, but one that's no less inhospitable, was the backdrop for Felicity Aston's record-breaking solo journey across the Antarctic in 2012. Only three people have crossed this land mass alone—two Norwegian men and Felicity. In her book, *Alone in Antarctica*, she describes a similar motivation to Robyn Davidson—"internal restlessness, a longing to leave everyone and everything behind." Being alone in such an empty, barren landscape brought about a state of isolation few of us can imagine.

Having trained as a physicist and meteorologist, Felicity was used to traveling to remote and unforgiving environments. In 2009 she led the largest women's team ever to ski to the South Pole. She was part of the first female team to cross Greenland, traversed Lake Baikal in Siberia, and was part of the first ever all-female team to complete the Polar Challenge, a 360-mile endurance race to the magnetic North Pole. But these were all group endeavors.

Felicity had ten years of experience in the polar regions and yet nothing could've prepared her for her biggest challenge—being alone. "The sense of absolute loneliness was instant, overwhelming, and completely crushing," she wrote. It was to be the biggest struggle of her trip. Not the extreme cold, the snow-covered crevasses, which nearly sent her plunging to her death, not skiing through a polar white-out, not pitching a tent single-handedly in a blizzard, not having frozen knickers, not eating freeze-dried meals every day—just simply being alone.

Felicity describes being "the only living thing"—no birds, no grass, no footprints. "Even bacteria have a hard time surviving in this part of the world…the scale of the emptiness was almost too much to absorb." She was frozen into her face mask and clothes, pulling heavy sleds and waking up every morning believing that she couldn't get through another day. Like Robyn, she doubted herself from the very start. "The first thing that struck me was, 'Oh my goodness, I can't do this, I don't want to be here, I've made a terrible mistake,'" she wrote.

As with Robyn Davidson's trip, the stakes were ridiculously high. Being so far away from rescue means that one moment of carelessness would equal death. At times, with nothing but white to focus on, no sense of distance, dimension, or scale, she experienced hallucinations and extreme vertigo. Some days she skied for over ten hours and only covered three miles. Wonder and despair were never far away, sometimes even occurring at the same time. "The most valuable lessons and insight resulted from experiences which, at the time, felt like the most miserable lows," she wrote. The worst moments became the best because she saw the best in herself through them.

For me, the thought of skiing anywhere is frightening, let alone across an entire continent. Strapping two thin six-foot planks to your feet and careering off a mountain? No thanks. In fact, the only time I have ever done it (after zero lessons because, of course, I'd madly signed up for a skiing press trip even though I couldn't ski), I had a breakdown halfway down a gentle green run. So Felicity Aston careering and jolting around deadly cracks in the ice, battered by hurricane-speed winds, dragging 85-kilogram sledges behind her, is almost unfathomable to me.

A few weeks in, she decided that she couldn't take it any longer and that she was going to have to return home. But she somehow found herself getting out the tent to continue for another day, and then another. Like Robyn Davidson, Felicity Aston came to realize that the journey of an entire continent starts with a single step (or a ski, in her case) and that, "No one can tell you the way, you have to find it yourself—and the way is never clear until you step forward."

The woman who was picked up felt very different to the one who was dropped off by the same plane two months previously. "I now knew that, despite tears and madness and anxiety, I would—and could—endure." Over the course of her trip, Felicity realized that her only limit was her loneliness, but "none of this insight would have been possible had I not been on my own."

While Guy and I were blissfully island-hopping around Greece or snuggling in country pubs in the Cotswolds (I did tell you we were sickening), an opportunity popped into my inbox. I'd been approached by Airbnb about a sabbatical scheme they were running to take volunteers to Antarctica. They asked me if I wanted to be one of them.

Although many of the trips I get to go on feel like once-in-a-lifetime opportunities, with this one, it really did seem to be the case. It was a landscape so vast and unknown to me with some of the most spectacular scenery on the whole planet. There'd be penguins. I'd visit the actual South Pole. This was finally some "proper" intrepid travel, the kind I knew I loved best. This would be an unforgettable trip that would push me way out of my comfort zone. I'd see beauty that reduces the select few people who get to go there to tears. I knew it would be far and away the most demanding and fulfilling trip I'd ever done. I'd felt so inspired by Robyn and Felicity, and wanted to see if I could test myself in the same way. It should've been my dream assignment.

Of course I said yes but, in truth, I was dreading it. I'd be assisting an environmental scientist with her research into microplastic pollution in Antarctica. I'd need a full medical and fitness exam before I was accepted. I'd be spending months in southern Chile, training with scientists and adjusting to the altitude sickness. As more details of the trip became apparent I started to get, if you'll excuse the pun, extremely cold feet. I was in that early phase of falling in love with Guy where even a night apart feels like a punishment.

The sabbatical would be for at least six weeks, ideally three months. Days would be spent collecting ice samples. Nights would be spent in a tent. A tent, *in Antarctica*. As someone who can't even handle more than a couple of nights' camping in the United Kingdom, in the summer, and who often still has the central heating on in June, the reality of what I'd signed up for started to sink in.

I had a similar feeling in the run-up to my wedding—extreme anxiety, waking up in the middle of the night with dread sitting heavy in my stomach, slight queasiness whenever I thought about it. Unlike with my wedding, I decided to listen this time. The fact was, I didn't want to go on this trip. Why was I forcing myself? What did I have to prove?

Wracked with guilt, I eventually called up the lovely publicist from Airbnb and told her I didn't think I was going to come along after all. She completely understood—no doubt preferring to hear this now, as opposed to discovering a journalist needed to be airlifted out of Antarctica—and I immediately felt flooded with relief.

Although it felt like the right decision, I couldn't quite believe I'd turned this trip down. The old me would never have said no to any assignment, especially one as incredible as this, one which was guaranteed to make me the most exciting person at any future dinner party ever. But I realized that I no longer needed to travel to the southern axis of the Earth, because I'd already spent a long time out in the cold. I had become an expert at operating in survival mode, of feeling remote and cut off from everything, nurturing a reliance only on myself. I'd had my feelings in deep-freeze for as long as I could remember. For the first time ever, I felt like I was thawing out. I think therapists call this a breakthrough.

In a weird way, I had already experienced a taste of Antarctica and the Australian Outback just by reading about these women's incredible journeys. And I realized that we had already shared some of the same feelings. Like Felicity Aston, I'd pushed myself to my own limits and felt the exhausting physicality of true loneliness. Like Robyn Davidson, I'd left behind almost everything and everyone I knew. I hadn't broken any records or planted any flags, but I'd gone on a long journey from where I started and now it felt like I was where I needed to be. Sometimes it's only when you're with someone that you realize how alone you felt before.

So my trip to the white continent was on ice. But I was still packing. Guy and I had decided to move in together.

HOW TO HAVE AN ADVENTURE (WITHOUT GOING TO THE EDGES OF THE EARTH)

Just being on your own can make even the most run-of-the-mill trip feel high octane and adventurous. However, it can be hard to feel intrepid when you're surrounded by people wearing fanny packs and waving selfie sticks. We all want a holiday that will transform us and shake us out of our normal selves. But maybe we also like spas and sun loungers (#sorrynotsorry). Thankfully it is possible to get a taste of the unexplored world without going into extreme environments. As Robyn Davidson put it, "One can choose adventure in the most ordinary of circumstances. Adventure of the mind, or the spirit." So whether you're on the trip of a lifetime, a budget break on the coast, or a record-breaking polar expedition, there are plenty of ways to have an adventure.

1. Turn your phone off.
Or at least on to flight mode. It's quite simply mind altering. You'll be amazed at how different you feel when you're not connected to your home self, even if it's only for a few days. It's almost impossible at first. It can even feel frightening. Your brain longs to hear the ding of a message and bathe in the reassuring glow of a screen. You will grab your phone aimlessly before you

even realize that it's off or that it can't do anything fun. But soon you start to tune into your surroundings in a different way. By removing yourself from your everyday life and resisting the urge to call or email friends or post photos to social media, you can allow a different consciousness to emerge. It's powerful stuff.

2. Do something scary.

Not necessarily every day, as the saying goes, and it doesn't have to be a bungee jump or a skydive. It can be as simple as trying the local food—if you've been to the Philippines and tried *balut* (a fertilized egg) or eaten a durian ("vomit fruit") in Asia you'll know that this can feel plenty risky. It might be a deeply personal yet intense experience, like a ten-day Vipassana silent retreat. It could be doing something you'd never normally do at home, like joining in a line dance in Texas or doing karaoke in Japan. This is another brilliant side effect of solo travel. When you're in a place where no one knows you, it frees you from feeling self-conscious. I've danced a tarantella in a remote village square in Puglia. I've dressed in a purple plastic gown to steam my vagina in LA. Would I have done those things if anyone I knew back home was there? Hell no. Well, probably the last one but only because it made a funny story. And no, I didn't feel any different afterward. It was like squatting over a boiling kettle.

3. Get up early.

Reeeally early. It pains me to write this, and I wish that it wasn't the case because I am definitely not an early riser. But if you want to have peace and quiet and maintain the mystique of "discovering" a major tourist attraction, then the early bird gets the best photos. Just think how smug you'd feel if you could talk about seeing the Taj Mahal at sunrise with only a handful of monkeys. Then you can go back to bed afterward.

4.   Travel during a weird time.

In the travel industry it's called shoulder season and it's the time "in between"—aka when no one else wants to go to a place either because of the weather or because it's outside of school holidays. You can make massive savings this way, get to feel more intrepid, and experience the place like the locals see it. Though do bear in mind that some restaurants/museums/attractions will either have shorter opening hours or might be closed entirely during off-season. The peace and quiet might be just what you're craving but equally the lack of people might bother you. I've been to cities in shoulder season and felt like I was in *28 Days Later*. Also, if the weather is really terrible—either stifling can't-go-out heat or torrential rainy season—you might need a back-up plan, or at least a raincoat.

5.   Do a bit of research.

Often there are unspoiled gems just waiting to be discovered that are an incredibly short distance away from the tourist hotspots, you just need to do a bit of digging to find out about them or do a bit of pre-planning to get there. Look at blogs, websites, Instagram, or ask locals. Be prepared to travel out of your way and off the usual route to seek things out. Maybe hiring a kayak for the day is worth it if it means you get to stop off at secluded bays. Perhaps you can be bothered to get a permit for that national park in Tuscany that only lets in a certain number of visitors each day. Or choose a whole country where you know tourist numbers are limited, such as Bhutan.

6.   Get back to nature.

To paraphrase the Romantic poets, it is in the presence of nature that we feel small and awestruck and full of wonder. But you don't have to be in a desert or on a glacier to find the sublime.

Stay in a bothy (a little hut for hikers) in the Scottish highlands, and you might not see another soul for days. Even in the busiest city, it's possible to find pockets of wild beauty. Go for a dip in the ponds in Hampstead Heath in London. Head to Deer Park in Delhi to hang out with monkeys, peacocks, and (you guessed it!) deer.

7. Go to an island.

Islands always have a feeling of adventure about them. Maybe it's a Robinson Crusoe thing, or because they take extra effort to get to, but marooning yourself on a patch of land is a simple way to feel like you're getting away from it all. When I visited the Andaman Islands (an archipelago between India and Thailand) in 2011, I felt like I'd stumbled on my own personal paradise complete with swimming elephant. All the extra faff of a connecting flight and a special visa and a choppy boat ride was well worth it. Similarly Silver Island is a yoga retreat held on a private island in Greece with no electricity or Wi-Fi. I've been twice now, and I came back zen AF. If you thought Tasmania was remote, get yourself to nearby Maria Island, which has "no cars, no shops, no worries" and pretty much no people. The Solomon Islands, east of Papua New Guinea, is one of the least visited nations on Earth. But be warned: Islands are often as hard to leave as they are to get to.

CHAPTER TEN
Life in the Fast Lane

Traveling dangerously with Aloha Wanderwell

The key ingredient that reliably turns a trip into an adventure is danger. Moving in with Guy seemed quick by anyone's standards and not without an element of risk. Not least because I was starting to realize that sometimes the scariest thing of all isn't running away—it's staying put.

Guy and I found a house on a tree-lined street in Hackney in East London and moved in our things together. He turned a blind eye to the ever-expanding skyline of toiletries around the bath. I pretended it was normal to own five surfboards when you lived in London. With the summer light and the surfboards and the palm tree that we could see from our kitchen window, I felt like I'd found the freedom and space I'd craved from California, but closer to home. From the outside, it might have looked like safety and security and settling down, that I had turned my back on adventure. But to me it felt like the opposite.

The truly dangerous thing was to let go and feel truly vulnerable with Guy in a way I maybe hadn't with anyone before. Deep down, I think I had always known that Sam would never leave me. And besides,

if he did, I already had one foot out. I always held a part of me back just
in case and constantly had one eye on the nearest exit. No doubt this
was down to my mum leaving. I was always just waiting for the next
abandonment and perhaps my traveling addiction was all so that I could
be the one to leave first.

But, with Guy, I was going all in this time, and it felt like doing
a skydive without a parachute. Although we had agreed to forgive and
forget what had happened to break us up before, a big part of me was terrified that something like it would happen again. I also didn't know how
I'd respond to commitment. Last time I'd looked for that—by getting
married—it hadn't ended so well. But I tried to channel the fearlessness
of the women I was reading about, who lived on the edge and sought out
jeopardy wherever they could find it.

One of the most daring was the brilliantly named Aloha Wanderwell.
Known as "the world's most traveled girl," she was the first woman to
drive around the world. She raced across seventy-two countries from
1922 to 1927, clocking up over 380,000 miles and traversing six continents. Dressed in military jacket, jodhpurs, riding boots, and a helmet,
she was dubbed "the Amelia Earhart of the open road" by the newspapers
that reported her exploits.

Born Idris Welsh, Aloha was just sixteen and a student at a convent
school in France when she spotted a job advert in the *Paris Herald* for
"a good-looking, brainy young woman" willing to "forswear skirts" and
"rough it" in Asia and Africa for an unspecified expedition. "Be prepared,"
it said, "to learn to work before and behind a movie camera."

"Young as I was, I was sure I had the qualifications," she wrote in
her book *Call to Adventure!*. "There is one thing about being born to
the life adventurous. The more trouble one meets, the more one dares
oneself to get out of it." She'd already had a fairly adventurous time of
it before convent school, traveling solo from Canada to England at age
twelve after her stepfather was killed in the First World War and learning
to drive at fifteen. In 1922, she joined the Polish promoter "Captain"

Walter Wanderwell's round-the-world stunt race as a mechanic. Despite no specialist knowledge in that area, she appears to have picked it up, describing herself as "tall and strong…with a certain facility for doing things." Being six feet tall, beautiful, and blond, Aloha unsurprisingly became the face of the expedition, capturing her adventures in a series of movie travelogues.

I may not have raced around the world in a Ford Model T, but I'd spent a lot of my life rushing through everything. Desperate to leave home and go to university, I now feel like my school years were spent ticking off exams and getting the right grades—and little else. Then I was speeding through university with the sole aim of "getting a good career," whatever that meant. Even when I was working in a job I loved, I'd often have been there only a few months before I was planning how long I'd stay and what my next move should be.

I was always looking for the shortcut, the life hack, the quick fix. I think it was partly this impatience that convinced me that Sam and I had to "take things to the next level" and get married. I thought life was on a tight schedule, and I had to keep up with everyone else. I wanted to be making progress and being productive at all times. I was constantly chasing whatever was just over the horizon—the next life phase, the next trip, the next move. So slowing things down and enjoying the present, as I was now, actually felt much harder.

Aloha's adventures behind the wheel frequently took her into dangerous situations. She snuck into Mecca, a place where, even now, non-Muslims and women under forty-five are not permitted to enter. She swam with sharks in the Indian Ocean—"That was a bad moment," she records—and was taken hostage by bandits along the Great Wall of China. According to Aloha's memoirs, she apparently charmed her way out of captivity by teaching her guards how to construct a machine-gun nest and work an automatic weapon. In Cambodia she nearly had her leg amputated after a mosquito bite became infected. After entering the USSR and making her way to Vladivostok in 1924, the Soviet Army named her an honorary colonel for "being the first demoiselle to pilot a motorcar to Siberia."

When they got back to the United States, Aloha and Walter got married and had two children, but it wasn't long before the open road came calling and they unapologetically left their son and daughter in foster homes. Their second trip became an even bumpier ride. While stranded in Africa and stuck for fuel, they lived with a tribe for weeks, until they ingeniously made petrol and oil out of crushed bananas and elephant fat. Aloha learned to fly a seaplane, which she landed on an unchartered part of the Amazon river. Footage from Brazil became Aloha's only cinema-released film, the ominously titled *The River of Death*.

After South America, the Wanderwells became estranged and Walter went to live on a boat in California. In 1932, he was shot and killed under mysterious circumstances. No one was ever charged with his murder but, for a brief time, Aloha was a suspect. She used the publicity to promote her latest film. Aloha carried on traveling, both alone and with her next husband, and continued directing and appearing in films. "I love getting home," she told a reporter. "But then, I love starting out again on another tour. Somebody once said to me that you never really live unless you live dangerously, and while it may not be true of everybody, it certainly is true of me."

When I travel alone I don't seek out danger, but I do tend to find I am on higher alert. When there's no one to share the responsibility, you can feel like you're working overtime to take in your surroundings and watch your stuff and navigate your way. When all the pressure of staying safe falls on you, you can feel acutely aware of every action and decision and, although it might not feel as relaxing as when you're traveling with someone else, that's not necessarily a bad thing. When our senses are more open, memories tend to stick in the mind more clearly. There's a thrill to being alone and out in the world that feels just the right side of unsafe. I love the feeling that anything can happen and that it might well be risky. Now I was discovering that I could also get this feeling at home and with someone by my side. For Guy and I, the future felt wide open and the stakes huge.

"Risk is the salt and sugar of life," said veteran traveler Dame Freya Stark, and she experienced more than a sprinkling of it over the course of her long, adventure-packed years. As I researched more and more female explorers, I kept coming back to Freya Stark. Cyclist Dervla Murphy cites her as her inspiration; Gertrude Bell praised her for beating the men at their own game; and some of her twenty-two books often top the lists of the best travel writing.

Born to artist parents in France in 1893, from an early age Freya was no stranger to accidents and unexpected misfortunes. A few weeks before her thirteenth birthday, her long hair caught in a machine in her family carpet factory in Italy, and part of her scalp and right ear were torn off. She always wore her hair on one side, as well as flamboyant hats, to conceal the scar.

Like Aloha Wanderwell, Freya described her wanderlust as a type of mania: merely the sight of a good map filled her "with a certain madness." Aged thirty-four, she struck out on her own, hitching a ride on a cargo ship heading for Beirut. She ended up in some of the most dangerous and remote places in the world—traveling in secret and by night through Syria and Lebanon—and, like Gertrude Bell and Hester Stanhope, was instantly captivated. "I never imagined that my first sight of the desert would enslave me right away," she wrote.

By 1931 Freya went trekking into hazardous and terrorist-controlled parts of Iran, where no Westerner had been before, with only a single guide and barely any money. She found the not-very-pleasant-sounding "Valley of Assassins," which she describes as "forbidden" to visit because it was so dangerous for outsiders, where "wave after wave of people unnamed and unnumbered lose themselves…in unrecorded dimness of time."

When the French army caught up with her, she played the fool and claimed to have been misdirected by her Thomas Cook guide book. She had no qualms about using perceptions of femininity in order to get out of a tight spot. "The great and almost only comfort about being a woman is that one can always pretend to be more stupid than one is and no one is surprised," she said.

She battled malaria and dengue fever and nearly died of a ruptured appendix. She spent a month besieged in the British Embassy after a coup in Baghdad in 1941, enduring intense heat, sniper fire, and the distant thud of bombs. On her release, she celebrated by buying three new hats. "There are few sorrows through which a new dress or hat will not send a little gleam of pleasure, however furtive," she wrote. Nothing could stop her traveling.

She married aged fifty-four, but it only lasted five years as her husband turned out to be gay. "I had an idea that my proper work was to love and be loved but it isn't," she wrote to her publisher, "it is just to write books, so what is the point of not doing so?" Although she had no children she had many godchildren in many countries and loved to take trips with them. She rode her donkey as she rode through life "with a calm eye for accidents and a taste for enjoyment in the meantime."

I tried to adopt a Freya Stark–esque aura of calm when I arranged to meet up with Sam again, after several months of no contact. I'd been happy to hear through friends that he had a new girlfriend but still felt nervous to tell him that I was now living with Guy. I'd come to dread these meetings that were fraught with emotion and ended in tears—always mine, sometimes both of ours.

We met up in a distinctly average Mexican restaurant in London to talk through the final stages of our divorce. Seeing an ex after a long time can feel like an uncanny experience. It reminds me of when you see an actor you know well from a TV show pop up in a totally different role.

The face is familiar and the old affection is there but everything else feels new and strange. Although in some ways it was nice to catch up, I think we both realized that we couldn't really be friends. Seeing our relationship written in such formal terms on legal documents made our marriage feel like it had happened to someone else. It was another life, and I had to leave it behind.

Yet there's still a part of me that, even now, can't quite believe that I'm not with Sam. We were together for so long and shaped each other in so many ways that the idea that he's no longer in my life still feels core-shakingly wrong. Occasionally Sam-shaped memories will grab me out of nowhere—fierce and stark—but rather than making me weep or want to block them out, I let them in, grateful to reminisce about what was such a long and happy chapter of my life. In my mind's eye, I sometimes see an alternative version of myself, one who stayed in their marriage, and I wonder what that woman is doing. Is she happy? Is it more of the same? But I know that ship has sailed. I chose a different route.

Guy and I were on a trip together in Cornwall when I realized that my Balinese bracelet had broken off, probably when we plunged into a freezing cold tidal pool in Bude. I was sad to say goodbye—it felt like the last threads (literally) of my single life—but also surprised that it had lasted so long. Losing it made me think about how much had changed since I first put it on. Two weeks later, I found out I was pregnant.

HOW TO SURVIVE SOLO TRAVEL WHEN THINGS AREN'T GOING WELL

Traveling by yourself can feel like a roller coaster. There's no one to share the intense highs with—like discovering a delicious new ice cream flavor for the first time—and also no one to share the crushing lows with, either. Something that on a trip with a partner or friend would become an in-joke you'd laugh about almost instantly, like missing the last bus or finding a pube in the sink of your Airbnb, can tip you over the edge into a full-scale, I'm-booking-the-next-flight-home meltdown when you're solo. So here are some tips to get you through the dark days and the low moments, and there will always be a few…

1. Follow the HALT rule.

Apparently this is something they teach alcoholics who want to reach for a drink. Are you Hungry, Angry, Lonely, or Tired right now? If so, don't make any decisions. The slight problem with this technique for me is that I am usually always at least one of those four things. However, as anyone who has screamed at their partner while trying to find that tucked-away restaurant recommended in the guidebook, "hanger" (hungry anger) can make you a very different (usually far more horrible) person.

And ditto those other tricky emotional states. Hungry and tired are normally pretty easily rectified. Anger is one that either needs to be unleashed or eases up with time. As for lonely, this brings me to point number two...

2. Find some people.

It doesn't matter if you don't talk to them, sometimes just being around other human bodies can help. Which is why the cinema—dark, silent, cozy—can be such a tonic. It is better if you talk to the other bodies, though. A problem shared and all that. So head for your hotel/hostel bar, the nearest bus stop, or the supermarket queue and get chatting to someone. Tell them all about your missed bus/stolen passport/homesickness. Traveling woes are universal. Chances are they'll have been through something similar. They might even have some good advice. If you're too sad and stressed out to even leave your hotel room (it happens)—FaceTime is your friend. Hopefully not literally, and you do actually have a mate you can call. If not... um...Siri?

3. Focus on the now.

This is a trick I learned from Sarah Wilson and her brilliant book *First, We Make the Beast Beautiful*, which is all about anxiety and how it can actually be beneficial. When you feel anxious, ask yourself, *What problem do I have right now?* Not tomorrow or next year or even five minutes from now. Your head will try and jump fifteen minutes ahead but keep it in the moment and you usually realize the worry has magically gone away. Obviously when you're traveling there's a chance that you are actually being chased by rabid dogs or have a rat sharing your toilet, or something's going on that you do need to act on now but, more often than not, by just being in the moment, the worries melt away.

Because the moment is all we have (and other things white men with dreadlocks will tell you on a beach in Thailand).

4.   Depend on the kindness of strangers.

Not to get all Blanche Du Bois on you, but the majority of people will go out of their way to help you if you ask them. The owner of the Airbnb I stayed in once in Mexico went out to buy me a sandwich, a Gatorade, and some hardcore Mexican medication when I asked him for a late checkout because I had a stomach upset. I met an American woman in my hotel lobby in the Himalayas who made me a hot water bottle using an empty Sprite bottle and a very hot tap because I looked cold (I have a cold resting face, similar to the bitchy kind, just more blue). A taxi driver in India took me where I needed to go for nothing because I'd had all my money stolen. Whether it's watching your luggage while you go to the toilet (the bane of a solo traveler's life) or simply holding your place in a queue, people like to help other people, you just have to ask them. The asking part is key though, and it's something I struggled with for a long time. But asking for help when you need it takes courage and brings crazy rewards. Like Gatorade, which is clearly revolting and yet such a fun color.

5.   Bring a taste of home.

I like to travel with a little something that reminds me of who I am, where I'm from, and the fact that I am going back there at some point. For me, it's Earl Gray tea bags. If I can't find any hot water, I have been known to sniff them. For a particularly fancy travel writer I know, it's a mini Diptyque candle that she lights to make even the most average hotel room feel five star. My Geordie friend Rachel, who now lives in LA, carries malt vinegar in her handbag at all times. If you haven't brought anything with you, playing a song, or watching a familiar TV show (hello, *Friends*) can be just as comforting.

6.   Embrace the new.

Yes, the place you're staying doesn't have a bath, and you really love baths. Okay, so the surfing course you booked yourself on and thought was going to be a week of you falling in love with a hunky instructor turns out to be run by an eighty-five-year-old man named Hector who speaks no English and possibly doesn't even know how to swim, let alone surf. Yes, it's infuriating that you don't know how to work the oven in your Mexico City Airbnb so you have to cook an entire Christmas dinner in a frying pan. But it wouldn't be an adventure if everything went to plan and was exactly how you liked it. It would be a trip to a shopping center. It would be a fantasy. As many an Instagram quote has told us, your comfort zone is nice and all, but nothing interesting happens there. Soon you'll be home and can have all the baths and *Point Break*-watching and roast dinners you can get your hands on—and these things will be all the sweeter for it. So for now just relish the novelty factor and the chance to do something different. You might even surprise yourself and like it. Also, you'll be dining out on that surf instructor story for years.

7.   Designate a worry time.

This mindfulness technique can be practiced anywhere, any time, but it's especially good when you're traveling solo. When you're alone, it's very easy for thoughts to spiral into a place that's not rational or helpful. So allocate yourself ten minutes a day to worry about things. Push any and every negative thought that doesn't need immediate attention into this doom slot and then get on and enjoy the rest of the day. You might even find that by the time your worry session rolls around, you can't even remember what it was you felt anxious about in the first place.

8.   Listen to audiobooks.

Listening to stories and podcasts can feel like you're traveling with a friend in your ears. They're good for times when you would normally read a book, but you're too tired/travel sick/ turning on a light will invite a deluge of mosquitoes who will treat you like an all-you-can-eat buffet. When you feel unwell or stressed, there's nothing more soothing than someone reading you a story to take your mind off what ails you, even if it's just for a moment. I feel like I am close friends with Michelle Obama, having listened to her entire autobiography on one hellish plane journey to Australia. I have been pulled through many a dark night of the soul by Alan Bennett reading *Winnie-the-Pooh*. Thanks for that, Alan.

CHAPTER ELEVEN
Destination Unknown

Finding yourself by getting lost with Virginia Woolf

To say I was shocked is an understatement. Completely freaked out would be another way of putting it. Although I'd come off the pill and Guy and I had talked about our hope of becoming parents soon, getting pregnant was supposed to take months, maybe even a year or more, not happen right away. There were so many things I wanted to do before having a baby. I wanted to do the Inca Trail! I needed to go to sex clubs in Berlin! I hadn't written a novel or mastered a headstand or learned to avoid using unnecessary exclamation marks! And now I was going to be a mother?

I'm well aware of how annoying this must sound to anyone who has struggled (or is currently struggling) to get pregnant, but there is nonetheless something discombobulating about it happening instantly. It just isn't what you're told to expect, especially at thirty-five. As Oscar Wilde said, "There are two tragedies in life, not getting what you want and getting it." I think a lot of women who have never been pregnant before and (if they're honest) haven't always been that careful wonder if it's even possible for them. There's a nagging feeling that maybe you've been infertile all along or that you really have left it too late. But the pregnancy test (and the ones I did after that just to make sure) were unequivocal. We were having a baby.

Guy and I were both pretty overwhelmed at first, but for him it quickly turned to excitement, whereas I still felt terrified. We'd only been back together for six months—was this another example of me rushing through everything to get to some imaginary finish line? I'd put big adventures like Antarctica on hold, were they now on hold forever? Would I still be able to enjoy a life in which I'd swapped jet-setting around the world at the drop of a cocktail umbrella for changing nappies and watching *Frozen* on repeat? What kind of mother would I be when I didn't even have a relationship with my own?

Early pregnancy felt strangely familiar because it was just like traveling. I felt a bit weird, was constantly tired, and food tasted strange. Those first kicks and flips in my stomach were like the feeling I got from turbulence on a plane. And just like when I'm traveling, I felt very underprepared and without a plan. But, as I eventually reminded myself, once I'd stopped staring into space doing the scream-face emoji, there's something wonderful about not having a plan.

Sometime previously, I'd realized that traveling alone without too much of an itinerary and nothing set in stone is the best way to see a new place. It puts you in a state of keeping your eyes open, listening, following your curiosity, asking questions, sniffing around, remaining open, trusting that things will work out. I needed to apply that philosophy to my whole life now, not just my trips. I was nervous and excited because I knew that my destination was going to be a massive culture shock. Motherhood is another country; they do things differently there.

When we go away, we often invest so much time and money and energy that we want a measurably memorable experience, something to prove it was all worth it, not realizing that it's the unpredictable, unexpected events that are often the most defining. We like to think that life,

and our travels, are linear, with goals we can move toward and a prede-
termined route we can follow, but of course they're not. When we rip up
the map and abandon the to-do list, we realize that plans are pointless
and instead we make space for chance. We often find something we don't
even know that we've been looking for.

Nearly all the women in this book speak of achieving remarkable
feats by being in the moment and not thinking beyond the end of each
day, the next step, the next mile. Elspeth Beard put it brilliantly when she
told me: "I think people overplan trips now. You need that element of
the unknown and that uncertainty to make it exciting. It's the ability to
get lost that's so important. When the unexpected happens that's when
the adventure starts."

One of the easiest ways to travel without a plan was once the only
way of getting around—going for a walk. It's easy to dismiss walking as,
well, a bit pedestrian, but before trains and planes and bicycles, those who
couldn't afford a horse really only had their own two feet to transport
them. And, if you were a woman, it was (and in many places, still is) often
considered risky to go out walking alone. For women, being a "street-
walker" meant prostitution. Many of the words we use when talking
about travel, particularly walking, have negative and sexual connotations.
Think of the "wandering eye," "straying," "that's why the lady is a tramp."
As late as 1895—when Nellie Bly and Annie Londonderry were making
their daring excursions around the globe—a woman found walking alone
on the street at night in New York could be arrested as a prostitute.

Although the figure of the flaneur in nineteenth-century Paris popu-
larized the notion of going for leisurely long walks in the city without a
destination in mind, this was a very masculine pastime. Women were, for
the most part, denied the pleasure of drifting aimlessly and if there was
a female "flaneuse," her stroll through the streets was only legitimized by
consumerism, as the rise of shopping arcades allowed women a safe arena
to stroll alone. Shops were seen as a liminal space between inside and out,
just as the hotel functioned as a halfway house between home and away.

Virginia Woolf was a big fan of solitary, aimless walks under the pretext of a shopping trip. In her essay "Street Haunting: A London Adventure" she describes taking a stroll to buy a pencil and "under the cover of this excuse" loses herself in the "greatest pleasure...rambling the streets of London." "As we step out of the house on a fine evening between four and six," she wrote, "we shed the self our friends know us by and become part of that vast republican army of anonymous trampers."

Solitary strolls like this weren't just escapism or a hobby for Virginia, they were essential. In her diaries, she often spoke about "walking off" her depression and dark thoughts. In 1934, she wrote: "I'm so ugly. So old. Well, don't think about it, and walk all over London; and see people and imagine their lives." For many of the characters in her books, walking is meditative, a way of escaping what Virginia called "the cotton wool of daily life."

In *Mrs. Dalloway*, Clarissa's decision to buy the flowers for her party herself and her "plunge" on to the pavement, sets off a journey of reflection, memory, and self-discovery. Through walking, this middle-aged woman can lose herself in the crowd and have an almost out-of-body experience, until her actual self seems "nothing, nothing at all." When she gets back home the spell is broken, she returns to her body and her insecurities, "suddenly shrivelled, aged, breastless."

The inspiration for the walks around London in *Mrs. Dalloway* were Virginia's earliest expeditions in her local park, Kensington Gardens, where she would encounter balloon sellers, skaters, nannies, park keepers and horse-drawn carriages on the road. Every summer, her family uprooted from 22 Hyde Park Gate to St. Ives in Cornwall. These long coastal rambles would later become the basis for her books *To the Lighthouse* and *The Waves*.

Sometimes, her characters don't even need to walk anywhere to achieve this liberation, they just need to be alone; a recognition by Virginia, perhaps, that women can't always find the time to walk because of the domestic duties expected of them. Mrs. Ramsay, the self-sacrificing wife and mother in *To the Lighthouse*, gets a similar sense of freedom from just being by herself in an armchair at the end of the day: "For now she need not think about anybody. She could be herself, by herself... To be silent; to be alone...and this self having shed its attachments was free for the strangest adventures...her horizon seemed to her limitless."

Reading this passage made me realize another important aspect of the solo trip for women. Because we still do the bulk of the emotional labor, a holiday with others is, for many women, not just a holiday. It also means being a part-time travel agent. In so many families and couples I know, it's always the woman who books the tickets and packs the clothes and researches the hotel. Therefore, a solo trip frees you from doing all this stuff for other people. You're forced to focus and tend only to your own needs, and absolutely no one else's, and suddenly, like Mrs. Ramsay, the possibilities are endless.

In 1914, Virginia and her husband Leonard moved from London to the suburbs of Richmond and later relocate farther out to Sussex. Although she initially missed the city (She famously said, "If it's a choice between Richmond and death I choose death."), when she was out of the suburbs and into the countryside she liked the effect her long daily rambles in the South Downs had on her thoughts and her writing. "I like to have space to spread my mind out in," she said.

Whether strolling through Hyde Park on a sunny day or shivering in a valley in Sussex, walking was part of the creative process for Virginia. She would often have imaginary conversations with her characters or people she knew while walking, giving new meaning to the phrase "rambling on." In her letters and diaries, she reported discovering many ordinary and extraordinary things on these solitary hikes through the Downs, from a snail getting lost under a leaf to an escaped circus monkey.

Although we don't necessarily think of Virginia Woolf as a travel writer, she did go farther afield than Bloomsbury, Cornwall, and Sussex, making numerous trips to Europe. In 1923, when she was forty-one and just becoming famous, she went on holiday to southern Spain. She was visiting her friend, the writer Gerald Brenan, in the remote village of Yegen. The pair went on long, hot walks and Virginia wrote about the sounds and sights of the Sierra Nevada and how it affected her way of thinking and writing. "The mind's contents break into short sentences. It is hot; the old man; the frying pan; the bottle of wine."

For Virginia Woolf and many other writers and philosophers, walking and thinking are inseparable. The mind wanders like the feet do—there are many possible paths to follow, multiple wrong turnings, the rhythm of the steps hammering out the sentences. I love Rebecca Solnit's assertion in her book *Wanderlust: A History of Walking* that the mind, like the feet, travels at three miles per hour. It's the slow pace of walking that makes it so powerful. "Walking allows us to be in our bodies and in the world without being made busy by them," writes Solnit. "It leaves us free to think without being wholly lost in our thoughts." Words to bear in mind as you plod along in the sleet at 4:00 a.m. from a house party to find the nearest night bus stop.

Walking can be done wherever you find yourself, with no equipment, and putting one foot in front of the other can, in the end, lead to some incredible feats of endurance. Women have done some remarkable things while walking. For example, Ffyona Campbell, who, in 1994, became the first woman to walk around the world. She covered 32,000 kilometers over eleven years and raised £180,000 for charity. Emma Gatewood was a sixty-seven-year-old grandmother who, in 1955, was the first woman to hike the famous Appalachian Trail solo. Mother to eleven children and a survivor of many years of domestic violence, Emma hiked in trainers and carried a shower curtain in lieu of a tent or sleeping bag.

I'm not sure what Virginia Woolf would've made of these feats. The higher purpose of them might not grant the same freedom that an aimless, solitary stroll does. There is a big difference between walking with a goal or destination in mind and walking for the sake of walking. Although the point of pilgrimages such as Egeria's is to get to specific holy places, flânerie is nonetheless spiritual in its own way—the very act of avoiding a destination can become meditative. Interestingly, the word *saunter* may have come from pilgrims in the Middle Ages, who wandered about the Holy Land, or "*à la Sainte Terre.*" Or it might be rooted in the words *sans terre*, without a home, without the ground—a nomad.

It took me a while to get my head around the idea of walking for the sake of walking. The first time I visited Sam's family in Suffolk and they suggested going for a walk I wanted to know, "To where? Like the corner shop?" To my city-dwelling mind, you only walked to get somewhere, not for the sheer enjoyment of it. Gradually I came to relish the sense of place that you can only get on foot, the difference in perspective, at home or abroad.

I've been lucky enough to go on a traditional safari a few times, and the most memorable experience wasn't a game drive, but a walking safari in Laikipia in northern Kenya. I spent three days camping in the bush with two Samburu tribesmen carrying spears and wearing checked blankets around their waists, as well as baseball caps and AK-47s. Hyraxes (giant brown guinea pigs) woke me up screeching at dawn, and we spent entire days just strolling through terrain that changed from dense thicket to parched earth and back again.

Although I didn't see any "big game," it was the smaller details, seen close up, that made an impression on me. The pincer ants which the Samburu use in place of stitches if they cut themselves; the elephant footprints bigger than a dinner plate; the fungus you could blow on, releasing spores to show you the way the wind was blowing. I felt fully present and at one with my surroundings in a way I don't think I would've experienced had I not been walking. That trip was proof that sometimes it really is best just to get out the car and find your feet, and I started going for long and aimless walks back home, too.

How you feel about walking depends so much on where you are, mentally and physically. I've had days where I could walk for hours, following my nose and my instincts, discovering cool things, eating delicious street food, feeling rejuvenated. Equally, I've had painful schleps where

I've dragged myself around, feeling out of place, unsafe and uncomfortable on foot. Yet I always know that on days when I feel a bit low or stressed, if I can motivate myself to go for a walk it unblocks something in my mind and I come back feeling lighter than when I set off. I'm not surprised that a study from Stanford University found that students who went for a walk came up with more original ideas than those seated at a desk.

How you feel about walking also depends on where you are geographically. In many places in the world there is still a sense that a woman on the street alone is "fair game." And those places might not be as far away as you think. A recent YouGov poll found that 49 percent of British women said they "always" or "often" felt unsafe when walking down an alleyway by themselves, and one in three regularly take steps to prevent themselves being sexually assaulted. Race, class, and gender play a huge part in whether loitering on the streets is an enjoyable activity or one that provokes suspicion, danger, and even arrest.

Some cities are simply better than others for the lone walker. I remember, on one of my first trips to LA, asking the concierge in my Hollywood hotel about the best way to walk to Beverly Hills. She looked at me with such genuine concern and shock that I might as well have asked her the best place to score crystal meth. "But it's only a few miles!" I told her, pointing to the map, and set off undeterred.

Two hours later, marooned on smog-choked, six-lane highway after getting lost in various industrial estates, I understood why. No one walks in LA. It is considered too spread out, the distances too vast to be covered on foot. There are too many cars and barely any pavements or pedestrian crossings. Instead, people go for "hikes" in the mountains, usually in very expensive workout gear following preordained trails and snapping selfies en route.

But on subsequent trips, I have come to love walking around LA. Once you've got your head around how dystopian it will feel at times, it's quite fun to be the only person strolling in a concrete jungle. It is much more challenging to aimlessly wander there, so there's a feeling of rebellion. Because as no one really walks anywhere in this city, just going for a stroll becomes an act of resistance. And because it's LA, you inevitably stumble upon some strangely profound (and also just strange) sights and sounds and people. Walking around Venice, I saw a dog in roller skates, an elderly woman hula-hooping, and a man selling ornaments made out of his own hair.

Strolling to lose yourself, and in the process to get a bit lost geographically, is an important skill to master. For the first few months after leaving my marriage, I had felt completely unmoored. The solid ground beneath my feet had slipped away, and I felt out of step with everyone I knew. Eventually I realized that there's nothing wrong with not being able to orientate yourself for a little while. In fact, it might even be essential. Saying we don't have a plan can feel passive, like you're just going with the flow and waiting for opportunities to fall into your lap, but I like to think that not having a plan means you're more actively present. You're not stressing about what "should" happen and instead making space for what "could" occur.

Getting lost occasionally is essential—not only while traveling but when traveling through life. The destination we think we want is often, when we actually get there, not what we need. I thought that I just had to get married to Sam and everything would fall into place. In fact, it was the opposite. When we make the focus of travel to get to one fixed point—the summit of Machu Picchu, that really cool flea market everyone told us about, the beach from *The Beach*—it's inevitably disappointing. Relinquishing control isn't easy, but it can be really powerful to just let stuff happen or, to paraphrase John Steinbeck, to not take a trip, but let a trip take us.

In *A Field Guide to Getting Lost*, Rebecca Solnit writes that she never does more than "flirt with getting [truly] lost...touching the edge of the unknown that sharpens the senses." From her own meandering, she deduces that mystery can itself act as a kind of compass and that,

"Never to get lost is not to live, not to know how to get lost brings you to destruction, and somewhere in the terra incognita in between lies a life of discovery." It's in our mistakes and deviations that we find a way in to who we really are. James Joyce, who wrote a pretty good book about a bloke going for a walk in Dublin, called errors "the portal to discovery."

I think that this truly random, lose-yourself meandering can only really be done alone. You need the freedom to go at your own pace, to be able to stop or move on or change direction on a whim and without a word. As Virginia Woolf proves, you don't need to go very far afield to have an adventure like this, just your nearest stationery shop. And it's important that Virginia's flânerie is driven by a pencil—as women began to renegotiate the city on their own terms, they were writing it, too.

Virginia Woolf's famous essay "A Room of One's Own" seems to be primarily remembered for encouraging women to acquire a home office if they want to write. But she explored the idea that a trip of one's own is equally important, too. She wonders what George Eliot's, Charlotte Brontë's, and Jane Austen's writing would've been like if they'd been able to travel. Indeed, Virginia recounts how she first got the idea for the essay while strolling through an "Oxbridge college" on an October morning—although she couldn't wander exactly where she pleased because women were not permitted to. "I hope that you will possess yourselves of enough money to travel and to idle," she wrote. "To…loiter at street corners… write books of travel and adventure."

Once I'd accepted the lack of plan and the loss of control that came with being pregnant, I was grateful that it had happened so quickly and felt thankful we didn't have months of slightly strained "trying" (a phrase I've always hated). Still I was already grieving my old life. The one where I could go for spontaneous saunas and cherry amaretto sours in Soho and eat whatever cheese I liked. But if I thought pregnancy was restrictive, I knew this was only the beginning phase of a total life overhaul.

Spontaneous saunas and cocktails—the spontaneous doing of anything, in fact—was going to be out for a while. It felt a bit like when you're close to dying in the first fifteen minutes of an exercise class and the instructor says, "Okay, great, I think we're warmed up now."

As a way of hanging on to my pre-baby life for a little bit longer, I resolved to pack as many trips as possible into the next nine months— Hawaii (surfing was out, but hiking volcanoes would be okay), Paris (no fun cheese or wine, but eclairs were still fair game), and Iceland (I'm pretty sure I couldn't heli-ski even when I wasn't pregnant, but everything else would be fine). And surely this was an excuse for a blow-the-budget beach holiday rebranded as a "babymoon"? I was determined to be the least-pregnant pregnant person ever and to carry on traveling just as much, if not more than before. Becoming a mum wasn't going to change me. I hoped that maybe I'd even be the type of carefree supermum who could strap on a baby and a backpack and breastfeed up the Amazon.

But as my body swelled and I started to fear being more than a few feet from a toilet, I knew deep down that having a child would change the way I traveled. But maybe it would be possible to set my own terms with regards to what it means to travel with a baby, and I might gain new freedoms in the process. Perhaps family holidays didn't have to be centered around tacky kids' clubs and fraught plane journeys.

Until the baby was actually here, I was determined not to let being pregnant curtail my travel plans. But it turned out that fate had other ideas.

HOW TO GET LOST

Can you plan to have no plan? There are a few ways to bring the element of chance into your holiday that don't involve waiting for someone to whisk you away on a surprise mini-getaway. Whisk yourself instead.

1. Let someone else plan your holiday.

If you're a bit of a control freak, then the thought of someone else booking your holiday might require a holiday to get over. But many travel companies are now catering to people seeking the element of surprise. Black Tomato offers a Get Lost package, which places the traveler in an unknown, off-grid destination and they have to navigate their way back to civilization. There's Srprs.Me, which only gives you a weather forecast and a time to be at the airport and then presents you with a scratch card to reveal your destination. There's also BRB, where subscribers pay a monthly fee to get sent on a random city break every few months. You only find out where you're going a few weeks before when they send you a postcard. I tried BRB and was sent to Seville, a city that I might not have got around to visiting on my own and where I ate my body weight in *jamon*. Maybe you even trust a travel-savvy friend to book your trip for you? Make sure it's one who knows you pretty well though and doesn't send you for a week on a beach in Marbella when your idea of holiday heaven is hiking Mount Snowdon.

2.   Ditch the maps.

It's far more amazing to stumble across Rome's Trevi Fountain than it is to navigate your way there with a blinking blue dot on your phone. When you allow a map to guide you to a place, it's a tourist destination; when you happen upon it by accident (via a couple of other cool things you weren't expecting to find), it feels like a miracle. Trust me, with the big sights you normally can't miss them. And if you really get stuck, you can always…

3.   Discover chat nav.

When you put your phone away and give the data-hungry map app a break, you're forced to find your route the old-fashioned way—asking people for directions. It took me a long time to get over my fear of interrupting strangers and asking them for help. But I've found that most people are not only more than happy to direct you if they can but they will often go out of their way to help you find what you're looking for—quite literally in the case of the people who will actually take you there. Often this leads to a little chat that, even if it's only the briefest of interactions, means you get to have an encounter with a local person outside of being served in a shop or restaurant.

4.   Choose a lost city.

No, I don't mean track down El Dorado. Go to an unmappable place that positively encourages getting lost so you have no other option but to go with the flow. The medina area of Marrakech is a place where even locals lose their bearings. Venice is famously impossible to navigate. Parts of Istanbul and Varanasi, India are similarly labyrinthine. Go somewhere chaotic enough and eventually even the most uptight travel planner will be forced to give up and drop their map in frustration, which is when the fun begins.

5. Put a pin in a map.

Yes, I do know people who have done this literally. The more modern version might be to use Skyscanner's "Anywhere" search function, EasyJet's "Lucky Trip" widget, or Google Earth's "I'm Feeling Lucky" button, all of which let you spin the travel roulette and see where you end up. When you get there, buy a local listings magazine, flick through it without looking, and stop on a random page. Whatever is on it is where you're going that night. That's how I ended up going to drag karaoke in Porto.

6. Practice the "yes method."

Spend your trip saying yes to every opportunity that crosses your path, whether it's talking to a chugger or accepting an invite to dinner. This is similar to the "end of the line" approach, where you get on a random bus/train/tram and just ride it until it won't go any farther. What a thrill!

7. Give yourself a down day.

For some people, part of the fun of a travel experience is researching and planning a trip beforehand. One friend told me that her boyfriend even decides what he's going to order in the restaurants at which he's already booked a table (psychopath). But try to give yourself at least one day with no obligations or reservations or plans to just explore and allow yourself to be spontaneous. No sights to see, no museums to visit, no bars to tick off the checklist, just go wherever you feel like the world is nudging you next. Yeah, you might miss that world-famous art gallery but that's not going anywhere, and you'll probably discover something a bit more interesting instead.

8. Pay attention.

When you're not hurrying around trying to follow a plan or a schedule (or buried in your phone), you can take pleasure in the minutiae instead. While out aimlessly walking around, I have seen things that I would definitely not have noticed if I had a destination to get to. I have happened upon a dog poo with a sparkler stuck into it in Paris. I once ended up in the middle of a political rally in Medellín in Colombia by following a cat I liked. I've encountered a witch doctor holding visiting hours in a shack in Malawi. I might well have walked right past those things if I hadn't been dawdling.

9. Don't actually get lost, lost.

If you're in a busy city then this is less of an issue, but if you're in a rural area make sure you're paying close attention to your surroundings—the route, the landmarks, the sun—so you can navigate your way back. Take a map and a phone just in case, and make sure you tell someone where you're headed and roughly what time you expect to be back. You don't want to get so lost that you need a search-and-rescue helicopter to come out and find you.

CHAPTER TWELVE
Home Sweet Roam

Having adventures in your own backyard with Nan Shepherd

Guy and I were on La Gomera, an unassuming speck of land in the Canary Islands, when I read the news about a new type of virus that was spreading through China. I thought about the people I'd met in Shanghai and hoped that they were safe. I felt pretty relieved that I wasn't in China, or planning on going there any time soon, but other than that I didn't give it much thought. Just over a month later the whole world was in lockdown.

One by one, all my planned trips were canceled or postponed indefinitely. Borders were closed, flights were grounded, cruise ships were quarantined, hotels were emptied, and major tourist attractions lay deserted. International travel came to a standstill—and then any travel at all. I went from feeling upset that I was no longer going to Hawaii to feeling grateful that I was still allowed to go to my local corner shop, all in the space of a few weeks.

When pregnant women were added to the list of those more "at risk" from COVID-19, I felt like I'd been placed under house arrest. They say there's no "perfect time" to have a baby, but slap bang in the middle of a global pandemic felt like the very worst time. The only twisted positive was that telling Sam that I was pregnant—a moment I'd been dreading—was now rather buried by the news of the impending apocalypse. I sent him an email, which seemed a bit cowardly, but I reasoned that sometimes it's preferable to process things in your own time. And besides, it could have been many months before we were able to meet in person. He responded with grace and kindness, reminding me all over again why I'd fallen in love with him in the first place and why he had been such a huge and defining part of my life for so many years.

The days felt longer and longer as the weeks of lockdown turned to months. I barely left the house, the groundhog days all rolled into one, with nothing to differentiate them other than a trip to the local park. My world had shrunk in a way that friends had talked about happening during early motherhood—so I tried to console myself that at least I was getting in some practice. It felt like all the color had been leeched out of life—there were no restaurants, no festivals, no bars, no fun, no holidays. Life felt stagnant, routine, and uninspiring. Whenever I'd felt unstimulated like this before, my default solution had always been to go on another trip. This time, I had to sit with those feelings instead. Travel had become such a big part of my identity that I wasn't sure who I was if I wasn't on the move.

The lockdown was the longest I'd stayed in London for well over a decade and I oscillated between feeling bored, numb, and anxious. A constant above all of these though was an overriding feeling of dread as the world I once loved to explore now appeared unrecognizable: the ever-rising death toll in Italy, the mass graves in New York, the nurses wearing bin bags instead of PPE in London.

Little by little, as a fortnight became a month, became six weeks, and then more, I started to realize that my solo travels had prepared me well for life under lockdown. When you travel alone, you have to be prepared to adjust your expectations to fit whatever you find, to be flexible and adaptable, to feel at home anywhere, to develop an acceptance of uncertainty. When you see the world by yourself, you have to be proactive and good at making your own routine because there's no one else to suggest things to do or bounce ideas off. All these skills now seemed essential for surviving lockdown with my sanity intact.

I began to find that home, and the two-mile radius around it, was opening up in a new way. I became obsessed with a family of fox cubs living in my next-door neighbor's garden and monitored them as excitedly as if I'd just caught sight of a leap of leopards on a game drive. Without cars and planes, birdsong seemed louder—or maybe I was just noticing it more. I started looking out for a bird with an electric blue streak on its tail (a Eurasian jay, I later found out) and two courting pigeons who chased each other around my balcony every day, sometimes fighting, sometimes nuzzling. I inhaled the jasmine growing on the corner of my street, marveled at bright purple rhododendron bushes in the local park, and took endless photos of the blossom on the trees and the wisteria blooms near my house. There was a whole world right on my doorstep; it had always been there, I'd just had to open my eyes and ears and nose to it.

With time (which I suddenly had *a lot* more of), I realized that I could explore my neighborhood with the same passion as I had previously explored new places. I began to see that I had been on a journey all along, and parts of it were as huge and terrifying and bewildering as any of my most adventurous trips abroad. The regions I had visited in my mind through my therapy sessions seemed like some of the most dangerous and impenetrable terrain I could think of. The journey of sexual exploration I went on after my marriage ended, which started with me road-testing high-tech sex toys for a magazine feature I was writing and ended with an evening of "pussy gazing" in an East London flat (it involved hand mirrors, low lighting, and saying thank you to my vagina) remapped the way I thought about my body and my relationship to sex. Travel doesn't have to mean a journey through space, it can be an inner change, too.

Life under lockdown pushed this newfound appreciation of home to its limit. But even before a wet market in Wuhan changed the world, I had started to appreciate the joys of being a tourist in your own neighborhood. After reading Virginia Woolf, I'd been trying to seek out less-traveled paths on familiar routes, going down curious alleyways I'd never noticed before and making a conscious effort to pay more attention to my surroundings. I had started to realize how lucky I was to live in London and be able to slurp noodles in a ramen bar, look at live giant snails at an African market stall, watch old Chinese ladies do tai chi in the park, and pick up Portuguese tarts—all on one street. But even in less multicultural surroundings, there are multitudes.

It was especially serendipitous that around the time the planet got grounded, I was reading about a travel writer who didn't travel very far at all. Nan Shepherd was born in West Cults, a tiny village near Aberdeen in 1893, and she died there in 1981. Through her many explorations of the Cairngorm mountains, she discovered a whole world. She spent hundreds of days exploring this mountain range in northeast Scotland on foot and in all seasons. She never married or had children or lived anywhere but the same house in her little Scottish village. She graduated from the University of Aberdeen and taught there for the next four decades, writing three novels set in rural communities in Scotland. Although she traveled widely—visiting Norway, France, Italy, Greece, and South Africa—she always came back to the Cairngorms, its foothills rising a few miles from her home.

Her book, *The Living Mountain*, written around 1945, describes her many journeys into the Cairngorms. Though it is now hailed as a "lost classic," the manuscript lay untouched in Nan's drawer for more than thirty years. It was then published in a very limited edition. It was reissued in 2008 and sold more than 90,000 copies, as well as being translated into a dozen languages. In 2016, Nan Shepherd became the first female author to feature on a British bank note when she was chosen to grace the Scottish five-pound note, looking ethereal yet steely with her braided hair and ornate headband.

I've never been to the Cairngorms (although it was one of the trips I had planned to take before the pandemic pressed pause on all my travels), but having read and then re-read the *Living Mountain* I now feel like I know this terrain well. Although I'm not a climber, and I haven't even heard of half of the plants Nan talks about, after reading her book of detailed, poetry-like prose, I feel like I've walked in these mountains, too. It's filled with the kind of acute detail that only comes from a lifetime of visiting the same spot and of "staying up for a while," as she puts it.

Nan effortlessly changes focus from the big and expansive moments—the swoop of a golden eagle, a plane crash—to the detailed and minute, "scarlet cups of lichen," "the down of a ptarmigan's breast feather," "small frogs jumping like tiddly-winks." She walks through the tops of clouds and swims in the deepest lochs. She studies animal tracks and learns to differentiate in the paw depressions between "a hare bounding, a hare trotting, a fox dragging his brush, grouse thick-footed." With the changing of the seasons, she notes huge shifts, but even hour-by-hour the landscape alters. In winter, she walks through "millions of sparkling sun spangles on the frosty snow;" in summer, the world is "flickering with midges by the hundred thousand." The mountain changes color "from opalescent milky-white to indigo" with something as subtle as a difference in the moisture in the air. It's a landscape as colorful as any tropical rainforest—mosses are "lush green," blaeberry leaves "flaming crimson."

We can't all have a mountain range in our back garden, but thanks to Nan Shepherd I've discovered that even the most urban and mundane environment rewards closer inspection. Let's face it, where I live in Hackney isn't the Cairngorms. I can't pretend that coming across discarded fried chicken bones, empty laughing gas canisters, and chewing-gum-spotted pavements is the same as listening to stags yodelling and foraging for wild berries. But during lockdown I came to appreciate London in a new way.

Strolling along the canal towpath in Hackney, I noticed cherry trees with actual cherries growing on them, bunches of rocket underfoot and the odd fig tree—sniffing their leaves took me back to Italy, or at least a very expensive scented candle I once owned. I was struck by how much free food there is all around us for the taking, which felt even more poignant given that queues for socially distanced supermarkets were at that point snaking around the block.

When I reached Hackney marshes, which could easily be in the middle of the countryside not the middle of Clapton, there were fields of elderflower, fennel, and even a "chicken of the woods"—a massive, rippling yellow mushroom growing in the trunk of an acacia tree. For perhaps the first time, I stopped to smell the roses. And the wild thyme and the lavender, too. By mid-afternoon, light was bouncing off the River Lea, bushes were vibrating with birds and bumble bees, and butterflies danced around, not seeming to care that they were within a wing's beat of the motorway as opposed to some bucolic idyll.

Nan has a deeply sensual relationship to the terrain she describes— "the feel of things, textures, surfaces, rough things like cones and bark, smooth things like stalks and feathers and pebbles rounded by water… the delicate tickle of a crawling caterpillar, the scratchiness of lichen"— which sometimes feels almost sexual: "to walk through long heather to feel its wetness on my naked legs." I tried to channel this as I walked through East London's marshes, feeling the gorse and the long grass tickle my calves, pausing to feel the bark on the trees and giving some a hug when I was sure no one was looking.

During lockdown, even the asphalt looked different. I took the time to read the chalk graffiti on the stone slabs—"Support the NHS," "Love = 2m apart"—and to idle over children's drawings of rainbows that seemed to have magically sprouted in every second window. Like Nan, I learned to tune into the sounds around me and to hear the birds chirping and the leaves rustling. It was so quiet that a passing double-decker bus, empty apart from the masked driver, felt more like a roar. With fewer people on the street, I had time and space to look up and I found myself spotting blue plaques I'd never noticed before, weather vanes, ornate clocks, garden gnomes.

On a rundown street near London Fields, behind the queue for an Iceland and next door to now-shut-up Empire Nail Bar, I spotted a plaque high up on the brickwork commemorating Celia Fiennes: "Traveler and diarist, 1662–1741, Lived in a house near this site from 1738, and died here." A google on my phone informed me that she was a pioneering solo adventurer who wrote a memoir about riding through every county in England on horseback. She began her side-saddle escapades for health reasons, visiting spa towns such as Bath and Epsom, and went on to provide a detailed description of England's industrial centers and social attitudes in the seventeenth century. I'm amazed that this intrepid and intriguing figure, who I had never heard of before, had lived right around the corner from me all this time and that the clue to her existence had been there all along.

I walked on past the silent building sites, their bright orange canvas awnings flapping in the breeze. Shop windows were dark, cafes closed, pubs boarded up. New cyclists wobbled past on red Boris Bikes, many almost as inappropriately dressed for a bike ride as a pre-bloomers Annie Londonderry. Before I knew it, I was at Monument, marveling at the columns and the cornicing of this Christopher Wren–designed memorial to the Great Fire of London, and how eerily empty and wrong the center of the city seemed without any people. I felt like I was in a post-apocalyptic future or the very last person to leave a party. Then I found myself on Fleet Street, a place I've never actually been before, but that has loomed large in my imagination as the spiritual home of journalism. The earliest newspaper offices and printing houses set up shop here in the eighteenth century, and Sweeney Todd supposedly did his demon barbering at number 152, next to St. Dunstan's church.

Down a cobbled side street, I peered over a small wrought-iron gate to see St. Bride's Church, its wedding cake–like white spire shooting defiantly out behind the clutter of gray buildings, its grand stained-glass windows revealing only a hint of the splendor inside. The church was also designed by Wren and, at the end of the street, I walked past St. Paul's

Cathedral—Wren's most famous project—its perfect dome looking stately and impervious to all the changes it has witnessed—fire, floods, bombings, plagues. I'd had no idea these places were so close to home, less than an hour's walk away.

Venturing farther along unfamiliar streets in my home city, I reached Virginia Woolf's old stomping ground of Bloomsbury and realized that I haven't been to the British Museum in over a decade. Of course it was all shuttered up, but it felt like a special privilege even just to see its grand exterior so stark and uncluttered by visitors. I thought of Mary Kingsley who brought back from Africa several "new" species of snake, fish, and insects that she donated to the museum, and they are probably still inside somewhere. Over the road is the London School of Tropical Medicine, an equally grand building I'd walked past many times before, but I'd never before noticed the gold mosquitoes that dot the ornate balconies. Seeing London so still and naked made it feel like *my* city, like it had all been laid out especially for me. Walking without a purpose or a crowd makes the world feel present and unhurried, like my very own personal playground.

Although at first, Nan Shepherd admitted that she approached the mountains "egocentrically" and "always made for the summits," over time she learned to appreciate the plateaus and to walk in them aimlessly: "Often the mountain gives itself most completely when I have no destination, when I reach nowhere in particular, but have gone out merely to be with the mountain as one visits a friend with no intention but to be with him."

It's clear from this quote why comparisons are often made between Nan Shepherd and Virginia Woolf (no doubt in part because they were both beautiful and bohemian modernist writers), but unlike Virginia, Nan didn't walk to lose herself—in fact, it was the opposite. Going into the mountains was a deeply physical experience that allowed her to find herself, to go deeper into her own being: "For as I penetrate more deeply into the mountain's life, I penetrate also into my own." She literally came to her senses. In the final sentences of her book, she wrote, "On the mountain...I am not out of myself, but in myself. I am. That is the final grace accorded from the mountain." *The Living Mountain* is a lesson in the rewards of a deep-dive into one place, as opposed to skimming the surface of several places. The mountain only revealed more layers of itself to Nan Shepherd. She could never truly know "the total mountain," like a "work of art [it is] perpetually new when one returns to it."

Similarly, in London, once I'd accepted being stuck with the same identical scenery, the repetition didn't reduce my pleasure but enhanced it. On every walk, I noticed something new: either the place had changed or I had changed. It was the opposite of how I'd been used to traveling up until that point, when I'd parachute into a destination with three days to race through the "Best of" list, or write a "Twenty-four hours in..." guide. I'd snap a picture of a sunset or a snow-capped mountain and move on.

Shepherd gave me a new perspective on what it means to travel and explore. She described walking into the mountains on a moonless night, with overcast skies and during a wartime blackout without a torch. She wrote of waking up on the mountainside ("no one knows the mountain completely who has not slept on it"). Sometimes she went up barefoot to really feel the textures under her feet. Occasionally she half-closed her eyes or adopted a different stance to get a different view: "Face away from what you look at, and bend with straddled legs till you see your world upside down. How new it has become!" I would love to be brave enough to try this on a busy street in London. And to be honest, as I got more and more pregnant, I would have loved to be able to bend down at all.

Nan Shepherd kept walking and foraging in the Cairngorms until the end of her life. Although she physically climbed the mountains until her eighties, in her final months, confined to a nursing home, her mind went on exploring. She hallucinated that her ward had been moved to a wood in Drumoak and that she could see Grampian place names around the room. When the body can no longer travel—whether that's due to old age, being heavily pregnant, or a global pandemic—we can go elsewhere in our imaginations.

As I am writing this, it is much too early to guess what travel will look like once restrictions lift. But there's no doubt that the world has handed us a hiatus and a golden opportunity to find solutions to problems such as over-tourism and climate change. I have known for a while that the planet can't support my jet-setting ways, but while I made a conscious effort to fly less, travel more sustainably, and offset my carbon footprint, there was always "one more trip" that I couldn't turn down.

I hope that this current situation means not the end of traveling and adventure, but an opportunity to reimagine what we want that to look like. I hope we might see the rise of a more meaningful, sustainable approach that benefits both travelers and the communities we visit. Imagine a world where instead of hopping on a plane and going somewhere for another tick-box weekend, we take our time, have an epic adventure, savor the journey, and enrich a place rather than deplete it. Until then, the pleasures of home not abroad will be rediscovered. Maybe we will travel less often and not so far in the future, but perhaps, like Nan Shepherd, we will travel deeper and better. Let's see.

HOW TO TRAVEL AT HOME

While I love the family who, during lockdown, recreated their canceled family holiday in their living room (complete with plane food dinners), there are other ways to travel without moving very much. Whether you've run out of holiday days, aren't physically able to travel, or are staring down the barrel of another staycation, you don't have to go far to feel far away. Immersing yourself in another culture doesn't have to mean going there in person, and "micro adventures" can be just as inspiring and satisfying (and they don't have any jet lag).

1. **Go on a blindfold walk.**
This is one thing you really can't do alone, and I know it sounds horribly cheesy, like something corporate types would do on a team-bonding day, but I once went on a blindfold walk as part of a spa retreat in Italy, and it reduced me to tears—the happy kind. At first it feels awful, relinquishing all control to someone else (in my case a total stranger). You feel weirdly self-conscious deprived of your sight, like everyone is staring at you. Then you're stop caring because you're touching, tasting, smelling, and hearing the world. Scent becomes more powerful. Touch feels incredibly

sensitive. I remember caressing olive branches, hearing the sound of insects humming, and tasting the air. At the end of the trail, I could smell a barnyard and I was guided to put my hand on something warm, furry, and alive. It was a horse. That's when I started weeping. Nan Shepherd believed that we can train our senses and that removing our most dominant one for a while like this is a good place to start. Obviously you need to be with someone you trust, but if you feel nervous about doing this in your local park, even just feeling out your front doorstep can be fun.

2. Camp out in your garden.

My friend's mum has a wee in her garden under the full moon every month because she says it makes her feel connected to nature. If you're lucky enough to have a garden, you don't have to use yours as a lunar-inspired toilet, but you could pitch a tent in it and take a mini-break in your own backyard. If you don't have any outdoor space, find a public park and have a picnic or make an outdoor cinema by setting up a projector and hanging a sheet between two chairs for a screen. Just being al fresco makes even everyday activities more exciting. This is why people put up with soggy sandwiches and warm Pimm's at picnics.

3. Get app happy.

Although we think of technology as disconnecting us from nature, if you download the right apps they can enhance the experience. If you live in London, enter your postcode into a website called Tree Talk, and it will generate a personalized map of the most interesting trees in your area. An app called Chirp! identifies birdsong and offers up human translations. Apps such as PlantSnap, Seek, or PictureThis help you identify what that cool flower or interesting plant is, like having a botanist in your pocket. Jeanne Baret would be proud.

4. Do some volunteering.

Spending a day—or even just a few hours—doing volunteering work near home is a great way to meet new local friends and perhaps see a side of the community that you'd never normally get to experience. If you don't know where to start, an organization such as Reach Volunteering hooks up people with opportunities in their area which match their skills.

5. Get virtual.

Some museums and galleries were already recording virtual tours for people who couldn't physically access their collections, but since the global lockdown this has stepped up a gear. Now, any time that you're in need of a change of scenery, you can do a 360-degree viewing of the Faroe Islands, take a tour through the Louvre, or—if you have a headset—get a virtual-reality peek inside the Vatican. A YouTube channel called Wind Walk Travel Videos lets you stroll down the Vegas Strip or travel around downtown Osaka. From your sofa, you can stream theatre productions in London, gawp at Russian ballet, or listen to the Stockholm Philharmonic Orchestra perform. Or just watch giant pandas going about their business thanks to the Smithsonian National Zoo's live panda-cam.

6. As seen on screen

Movies are a powerful way to make you see a place anew, whether it's the Mexico City of *Roma* or the Tuscany of *Call Me by Your Name*. Search online for films and TV shows that were filmed in or near your home town to make you see it in a whole new (often glossier) light.

7. Shop around the world.

I love souvenirs, especially practical things that I know I'll actually use at home, such as an ornate box of matches or a beautiful pair of gold scissors. Every time I use the latter I'm reminded of the little design shop in Gothenburg I bought them in. But you don't actually need to go to a place to bring a souvenir back home. Whether it's the signature candle of your favorite hotel that transports you every time you light it or stocking up on Korean beauty products without traveling to Seoul, online shopping is a good substitute for going there. If you're anything like me and are a sucker for a good hotel gift shop, check out The Hoxton and The Ritz, which both offer up their curated hotel boutiques online.

8. Enjoy world music.

There's nothing like music to take you places, which is why my travel playlists are filled with the country songs I heard in Arizona and the Sega music I first listened to in Mauritius. I also like to tune into radio stations from other countries. Streaming Fip Radio, I feel chic and Parisian, even if I'm only doing the laundry. I recently discovered Radio Garden, which is an audible Google Earth, allowing you to click and drag a 3-D globe around, and zoom in on radio stations around the world, whether it's a Ugandan pop station or a spot of Norwegian evangelical rock.

9. Taste adventure.

Food is such a big part of the travel experience for many of us, whether it's finding the perfect arancini in a Naples backstreet or the best *nasi goreng* in Indonesia. Luckily it's also one of the easiest aspects to recreate at home. Buy a cookbook from your favorite country, make your own Italian aperitivo, or order some wine from a vineyard you've visited or want to go to. During lockdown, many top chefs even started virtual cooking lessons, so you can knock up the dishes you love from some of the best restaurants in the world.

10. Start plotting the next trip.

If you can't go away imminently, it's a good time to book the type of holiday that requires a bit more organization than most. Destinations such as the Galapagos Islands or safaris often get booked up months (sometimes a year) in advance. Use this extra planning time to ensure future holidays will be more environmentally friendly than before, perhaps exploring the option of traveling entirely by train or donating to a charity such as Tree Sisters, which will offset the carbon of your flight by employing women to plant trees in deforested areas. Start bookmarking beautiful Airbnbs or making a Google Maps list of all the bookshops and bars you want to visit on your next holiday. A 2010 study found that just planning or anticipating a trip can make you happier than actually going on it.

DEPARTURES

When I first had the idea for this book, I was single and grieving the end of my marriage and everything I'd lost. Life as I then knew it had ended. Now I'm living with someone new and, by the time you read this, we'll have a baby. I guess this shows not only how long books take to write (way longer than making an actual human, it turns out) but also how quickly everything can change in a short space of time. At one point I felt like all was lost, that I had run out of time to have a family and that my life was over—I wish I could've known then that a new life was only just beginning. I'm the proof that sometimes when the things you hope for feel a long way off, they are actually only just around the next corner. All you have to do is keep going.

We like to use traveling metaphors and apply them to life—it's "a journey" after all—but when I look back on the place I was in when I started this book, I feel like my internal geography has shifted a few continents. I've realized that sometimes you have to turn your back on your destination to get there; sometimes you're furthest away from your goal when you feel closest. That often the only way around is the long one. That a place or destination is just as much about our perception of it as its reality. Maybe it wasn't a mistake, you just took a more interesting route to get there. As Gertrude Stein put it, there is no there, there. There is no end goal to be reached or summit to be conquered.

I know that without all the solo trips I've taken I would never have had the courage to leave my marriage, that in a way, they were like mini trial runs for what was to come next. And then, in the wake of my divorce, they became my salvation. I can't claim to have been even half as adventurous or intrepid as the women I've written about in this book. As an explorer, the only unchartered territory I've discovered has been within myself. Yet their stories of grit and determination gave me hope and propelled me to keep on moving forward at a time in my life when I felt like I was clinging on for survival. More than that, they showed me that there is more than one cookie-cutter version of womanhood to aspire to and many different ways to live a full life.

I could never have imagined that by the end of this book, I'd be living in a world where I was unable to travel farther than a few miles from my front door, that my idea of a big adventure would be going around the block, not going around the world. But the female writers I'd been researching for so many months provided a lifeline when I desperately needed one. Reading became a way of coping with confinement through armchair adventures. Through their stories, I was reminded of things that happened on my own travels, but I also traveled with them to places I've never been and might now never go.

From my bed, I went trekking through "the gaunt, leopard-colored lands" of Yemen in the company of Freya Stark. From my sofa, I could screech and stall around the Giza Pyramids in Egypt, getting into scrapes with Aloha Wanderwell, or career around India on a broken motorbike with Elspeth Beard. I realized that I didn't have to actually go to any of these places, I could be transported there through these women, via the power of their words. Just as it was when I was a child, reading became a way to find my way home—not an external place, but within myself.

These women's stories and experiences took on new meaning while living under the haze of a deadly virus. Stripping life of all but the essentials, as we all had to do under lockdown, is something Felicity Aston talks about on her expeditions. "By emptying life of all but the basic concerns—eating, sleeping, staying safe, you reveal the fundamental structure that underpins the whole and you discover what you hold most dear."

I haven't made it to Hawaii yet, but until I do I think Juanita Harrison would be pleased to know that I've been shopping for lunch in my local market and splashing in the surf in Wales. Was it Waikiki Beach? No. But I've certainly had a crash course in nursing the joys. Although I had no Yellow Brick Road or dancing munchkins, I've done a Dorothy. What I'd been looking for was in my own backyard all along; it just took a global pandemic and writing this book to recognize that.

Deciding who to feature in these pages was one of the hardest parts of writing *A Trip of One's Own*. There were many more women who I didn't have space to mention. Sometimes it felt like every time I delved into the life of one unsung female solo traveler, there'd be ten more behind her, begging to be written about. Many I discounted only because they traveled with others—and not alone—or because there was either not enough or too much material on them, but their stories are no less pioneering or inspiring.

I wish I'd had space for Gudrid Thorbjarnardóttir (better known as Gudrid the Far-traveled, understandably), a Viking woman who, sometime in the ninth century, visited Norway, Greenland, and Rome. According to the Icelandic sagas, she even sailed to North America in a longboat, which meant she beat Christopher Columbus's voyage to the New World by some 500 years.

I have a special place in my heart, if not this book, for Edwardian-era wild child Aimée Crocker, who amassed a huge collection of Buddha statues, pearls, tattoos, snakes, adopted children, husbands, and lovers. She escaped a poisoning in Hong Kong, an attempted murder by a knife-throwing assassin in Shanghai, where she had shacked up with a feudal Chinese warlord (as you do), and had "sensual encounters" with a boa constrictor in India and a lute in China. Maybe you had to be there. While accompanying a Bornean prince to Indonesia, she found herself the target of headhunters (not the recruitment kind) and would throw parties with a sixty-pound boa constrictor around her neck. Sorry, Britney, Aimée got there first.

Slightly more sedate but no less adventurous was the artist and botanist Marianne North, who in 1870 decided she'd had enough of painting English flowers and went around the world to paint them instead. She relished her "spinster" status for the freedom it gave her to travel. "Marriage? A terrible experiment," she once wrote. In 1882, an exhibition of her paintings opened at Kew Gardens and is still there today—the only permanent solo exhibition by a female artist in Britain.

I often think about the Belgian French explorer Alexandra David-Neél, who disguised herself as a beggar and Buddhist monk and became the first Western woman to meet the Dalai Lama when she snuck into Tibet in 1924, and who was forced to eat her frozen boot leather when she ran out of food. Snacking on shoes might have been more palatable if she'd run into the novelist Sybille Bedford, who took a rambling trip through Mexico in the 1940s with a traveling case stocked with cocktail glasses and a miniature pepper grinder.

For obvious reasons, I identified with Rosita Forbes, who in 1917, at twenty-seven, dumped her husband, pawned her wedding ring, and attempted to ride from South Africa to England alone on horseback before taking a detour to thirty other countries along the way. And I loved the sound of the pioneering trio consisting of the missionaries Mildred Cable, Francesca French, and Francesca's sister Evangeline, who spent thirteen years crossing the Gobi Desert in the 1920s.

There were so many brilliant female pilots that I wondered why Amelia Earhart gets all the credit and whether *aviatrix* might be my new favorite word. Take Jean Batten, who purloined money from a string of hopeful young men to fund her record-breaking flights in the 1930s, including the first ever direct solo flight from England to New Zealand. There was

Bessie Coleman, the first African American and Native American woman to get her pilot's license in 1921, who wowed crowds across the United States with her loop-the-loops and tail-spins. She hoped to start a flying school for women but died in a plane crash aged just thirty-four. When it came to living the high life, Beryl Markham chalked up some impressive achievements and glorious scandals. In 1936, she was the first woman to fly solo over the Atlantic against the prevailing winds and included an English prince and two famous authors among her many lovers.

Whether over land, sea, or sky, all of the travelers in this book subverted expectations of womanhood, pushed the frontiers of feminism forward, and reveal the complexity of the terrain we try to navigate as women, not just centuries ago but today. No wonder their traveling experiences still resonate—women continue to be victims of violence, are still judged by their bodies and what they do with them, and still have to dodge the same old questions, usually about why the lady traveler is not married or doesn't have children yet. Global pandemics aside, in many places in the world today women remain housebound, whether by law or custom or fear.

A solo adventure to a far-flung land won't be happening for me in the foreseeable future. Traveling with a baby will probably require more planning and luggage than a military campaign, and I'm sure that I'll be wistful for the days when I could wake up in the morning and stride out of my hotel with no bag and no plan and no place to go and the intoxicating feeling that anything could happen. I know that I will do that again, and hopefully it will be even richer for the waiting. I hope that this book has inspired you to go somewhere new or see somewhere familiar in a new way.

Thankfully, I have a lot of travel memories to draw on and when they run out I have these women's stories to dip back into and transport me. Whenever I'm craving escape, I'll remind myself of this quote from Freya Stark, whose words also began the book: "Good days are to be gathered like grapes, to be trodden and bottled into wine and kept for age to sip at ease beside the fire. If the traveler has vintaged well, she need trouble to wander no longer; the ruby moments glow in her glass at will." I hope she'll understand me changing the pronoun.

Like many women in this book, who had adventures right up until the end of their lives, I know that I'll never finish traveling. The global pandemic and the thought of impending motherhood has only fueled my wanderlust and made me appreciate time alone even more. I hope to carve out the space for at least one solo trip a year, no matter what else is happening in my life. I never intended for this book to be a journey from carefree independence to cozy domesticity. But when it comes to adventures, I have to accept that it won't be "just me" for a while. It will be just the three of us. For now, at least, the wandering is on pause. The journey is only just beginning.

READING GROUP GUIDE

1. How do you define "an adventure?" Was your last adventure local or abroad?

2. Kate admits that she's felt at times that "receiving male attention is the ultimate goal" of anything she does. Have you dealt with similar feelings? Do you think her intention to remain celibate on her journey through Israel and Palestine was an effective way around that habit?

3. Traveling solo, Kate misses having someone to share her humorous observations with or to take candid photos of her. How would you get around these drawbacks?

4. Kate recommends training yourself to be more comfortable in your own company by taking yourself on small solo adventures locally. Have you tried this? What was your experience like?

5. Nellie Bly was certain that audacity and will could get

around most obstacles to a last-minute trip. What scares you most about deciding to do something spontaneous?

6. Martha Gellhorn observes that "the only aspect of our travels that is guaranteed to hold an audience is disaster." Do you have a favorite "horror journey" story? Why do we have such fun telling stories about travel mishaps?

7. How does travel break up your routines? Why is it easier to do some things (like exercise, for example) when you're far from home?

8. Elspeth Beard had trouble readjusting to her normal life after her global motorcycle trip. How does traveling affect your relationship with home, and how are those effects magnified by traveling solo?

9. Taking a walk can allow you to really experience a place, local or abroad. Which of your five senses do you feel most keenly when you walk without a purpose? What is the most unexpected thing you can find in your neighborhood this way?

10. What travel tip will you be taking with you on your next adventure?

A CONVERSATION WITH THE AUTHOR

Since you travel for work, what do you do differently to go on vacation? Is your approach to a new place different when you're there purely for yourself?

It can be really difficult to switch off from travel-writer mode. Even if I'm not writing about a hotel, I can't help but mentally rate the mattress or try to think up something funny to say about the spa. It can be a bit annoying, but it's also not the worst thing in the world. The traits I need as a travel journalist—being curious, discerning, inquisitive—also make for a good holidaymaker. That said, sometimes you do just want to lie facedown in a sun-lounger for a week, which I would never get away with on assignment. I think all my travel experiences get filed away for potential future features. As Nora Ephron said, "Everything is copy." She was talking about heartbreak, but it also applies to holidays!

Do any of your travels stand out as your favorite? Why does that trip hold a special place in your mind?

Some trips stand out because of the company or the time in my life rather than the place. But one destination that I will never forget is Iceland. I lay in a milky lagoon watching the Northern Lights flicker across the sky and ate "geyser bread" that had been baked in the geothermal warmth under the ground, washed down with Brennivin, an Icelandic schnapps that translates as "burning wine." Nearly all of my most memorable travel experiences had to do with nature in some way. That and food and/or drink!

Do you research a new place before traveling there for the first time? Does your research ever change your first impressions of a place?

I try not to over-research (or at least that's my excuse for being a bit of a lazy planner!). I like to know a little bit of history and a few phrases of the local language. I will ask travelers I trust for their tips. I try to get the balance right between anticipation and overpreparation. Too much research can lead to preconceptions. I like to do most of my fact-finding on the ground. That way you get to have a fresh first impression, but you can also back it up with a bit of knowledge and context. And I have never understood why people will trust online reviews on sites like TripAdvisor over friends or locals.

Though their accomplishments and writings are crucial pieces of history, some of the women whose trips you followed also draw modern attention for their racist, exoticized accounts of places outside Europe and America. How do you approach their work?

Yes, it did feel quite jarring at times to come across the casual racism in some of these women's writing. I didn't want to gloss over it or make excuses for it. But I did try to see it in context and work out what it could teach me about the time they were living in. Having said that, I'm aware that I am approaching these texts from a position of privilege, and perhaps readers of color would have had a different reaction to them.

Unlike Egeria or the other historical figures, you were able to meet Elspeth Beard in person. Did that change the way you incorporated her story into *A Trip of One's Own*?

Yes, being able to meet Elspeth meant it could be more of a dialogue, which was great. She was every inch the intrepid biker I had imagined her to be. It also made me wish I could've met the other figures in the book too. Imagine taking Emily Hahn for a cocktail or going for ice cream with Juanita Harrison!

Are there any locales that you're still waiting to cross off your travel bucket list?

So many! I can't believe I've still not made it to Japan. And having only recently passed my driving test, a California road trip is very much on the agenda. I'm really keen to take Blake to some of the places I've loved so that she can experience them too. I'm writing this from Mauritius—the first time I've taken her out of Europe—and watching her face light up as she paddled in the Indian Ocean was a delight.

BIBLIOGRAPHY

ARRIVALS

Sapiens by Yuval Noah Harari

CHAPTER ONE—SISTER ACT

Egeria's Travels by John Wilkinson
The Pilgrimage of Egeria by Anne McGowan and
    Paul F. Bradshaw

CHAPTER TWO—SLEEPLESS IN SHANGHAI

China to Me: A Partial Autobiography by Emily Hahn
No Hurry to Get Home by Emily Hahn
Recipes by Emily Hahn
*Nobody Said Not to Go: The Life, Loves,
    and Adventures of Emily Hahn* by Ken Cuthbertson

CHAPTER THREE—ON ASSIGNMENT

Around the World in Seventy-Two Days and Other Writings
    by Nellie Bly
Travels with Myself and Another by Martha Gellhorn

CHAPTER FOUR—THE WOMEN'S MOVEMENT

Around the World on Two Wheels: Annie Londonderry's
    Extraordinary Ride by Peter Zheutlin
I Have Been Young by Helena Swanwick
The Memoirs of Ethel Smyth by Ethel Smyth

CHAPTER FIVE—A MAN'S WORLD?

The Discovery of Jeanne Baret: A Story of Science, the High Seas,
    and the First Woman to Circumnavigate the Globe
    by Glynis Ridley
Isabelle: The Life of Isabelle Eberhardt by Annette Kobak
The Nomad: The Diaries of Isabelle Eberhardt
    edited by Elizabeth Kershaw
Through Persia in Disguise by Sarah Hobson

CHAPTER SIX—FIGHT OR FLIGHT?

Lone Rider by Elspeth Beard
Eat, Pray, Love: One Woman's Search for Everything
    by Elizabeth Gilbert
Wild: A Journey from Lost to Found by Cheryl Strayed
Paradise Lost by John Milton
Angels in America by Tony Kushner
The White Album by Joan Didion

CHAPTER SEVEN—THE TRAVELING CURE

A Lady's Life in the Rocky Mountains by Isabella Lucy Bird
Travels of Lady Hester Stanhope by Charles Lewis Meryon
*Star of the Morning: The Extraordinary Life
    of Lady Hester Stanhope* by Kirsten Ellis
A Woman in Arabia: The Writings of the Queen of the Desert
    by Gertrude Bell
The Letters of Gertrude Bell Volumes I and II by Gertrude Bell
Travels in West Africa by Mary H. Kingsley
Hints to Lady Travelers by Lillias Campbell Davidson

CHAPTER EIGHT—FOOTLOOSE AND FANCY-FREE

My Great, Wide, Beautiful World by Juanita Harrison

CHAPTER NINE—THE GREAT ESCAPE

Tracks by Robyn Davidson
Alone in Antarctica by Felicity Aston

CHAPTER TEN—LIFE IN THE FAST LANE

Call to Adventure! by Aloha Wanderwell
The Valleys of the Assassins and Other Persian Travels
    by Freya Stark
Passionate Nomad: The Life of Freya Stark
    by Jane Fletcher Geniesse
First, We Make the Beast Beautiful by Sarah Wilson

CHAPTER ELEVEN—DESTINATION UNKNOWN

Street Haunting: A London Adventure by Virginia Woolf
Mrs. Dalloway by Virginia Woolf
The Waves by Virginia Woolf
To the Lighthouse by Virginia Woolf
A Room of One's Own by Virginia Woolf
The Diary of Virginia Woolf, Vol. 5: 1936–41 ed.
    Anne Olivier Bell
Travels with Virginia Woolf by Jan Morris
Wanderlust: A History of Walking by Rebecca Solnit
A Field Guide To Getting Lost by Rebecca Solnit

CHAPTER TWELVE—HOME SWEET ROAM

The Living Mountain by Nan Shepherd

ACKNOWLEDGMENTS

To Guy Felix Thompson, who has flipped my world upside down a few times over. Thank you for making every day feel like an adventure.

To my sister Anna, my source of support and encouragement in all things. And to my father, Peter, who first read me the stories that opened up my world.

To Francesca Zampi, thank you for believing in me, and this idea, right from its beginnings in a café in Notting Hill. And to Katie-Jane Sullivan for your enthusiasm. To Susannah Otter, editor extraordinaire, who made this book so much better and just got it from the word go.

To all my brilliant friends who pieced me back together again. Emma Ledger, my partner in (actual) crimes, ponds, and fellow Fun House twin—see, you get a whole line! Sarah Chadfield, my first reader, WhatsApp therapist, and voice of reason. Ella Bowman, who came into my life right when I needed her most. Jessica Butler, Nicola Pastore, and Phillipa Law, who have traveled by my side for two decades and counting.

To Alice-Azania Jarvis, Ellie O'Mahoney, Catherine Bennion-Pedley, Olivia Squire, and every editor who has sent me on assignment to somewhere exciting. Thank you for letting me explore the world and call it work.

To Josh Bullock, for providing me with a room of one's own to write (and have dance parties), from New Cross to the Dordogne. To Rachel Richardson, for letting me check in to the Heartbreak Hotel in LA.

To Christine, Esmé, Ross, Eloise, Dave, Elodie, Tansy, and Spike. Thank you for making me feel so at home.

To Sam, you have shaped the person I am today. And to Sara, J, Rachel, and Josh for showing me what a family could be.

To all the solo female travelers past, present, and future. Thank you for inspiring me with your adventures. To paraphrase one woman in a bar in Bangkok…let's go and get our share.

ABOUT THE AUTHOR

Kate Wills is a freelance travel and features writer for *Vogue,* the *Times,* the *Guardian, ES Magazine, Grazia,* the *Telegraph, Elle,* and many more. She has a weekly column in *Fabulous* and regularly appears as a commentator on Sky News, National Geographic Channel, and BBC radio. Kate is also the host of her podcast *Ticket for One.*